Final Voyage
of the
THETIS

Immigration and Smuggling in the 19th Century

Final Voyage of the Thetis

Published 2024 by Paul O'Dowd
Copyright © Paul O'Dowd
ISBN: 978-1-916544-40-6

All rights reserved. No part of this publication may be reproduced or transmitted in any form or by any means, electronic or mechanical, including photography, recording, or any information storage or retrieval system without permission in writing from Paul O'Dowd. The book is sold subject to the condition that it shall not, by way of trade or otherwise, be lent, copied, altered, resold or otherwise circulated without Paul O'Dowd's prior consent.

Map Illustrations by Parvathi Venkitaraman

Publishing Information
Design & publishing services
provided by JM Agency

www.jm.agency
Kerry, Ireland

Final Voyage
of the
THETIS
Immigration and Smuggling in the 19th Century

Paul O'Dowd

This book is dedicated to my grandsons Lucas and Antonio.

Contents

Introduction	09
The Shannon Estuary	12
The Thetis	26
Francis Spaight (1790 -1861)	47
Why Quebec?	79
The Journey down the Shannon Estuary	100
Crossing the Atlantic Ocean and the Gulf of St Lawrence	138
The Return Journeys July and October 1834	168
The Native (February 1843)	192
Epilogue	202
References	205
About the Author	207
Acknowledgements	208

Introduction

I have often sailed the Shannon Estuary from Foynes in Co. Limerick to Loop Head in Co.Clare. I have experienced enough sailing in the Atlantic and the Mediterranean to be able to appreciate the vastness and the solitude of the seas. Being interested in sailing, I was curious at the lack of information available about a shipwreck on a beach in North Kerry. The wreck is all that remains of a ship called *the Thetis*.

The wreck of the Thetis, as it stands today in Beale Beach. The official record reads 'Wreck no. W05.943, Latitude 52.577 Longitude 9.633.

It seemed that there was something not quite right about the accepted story of the wrecking of *the Thetis*. During the years of the pandemic, I decided to try and find out how the ship ended up on Beale Beach. I have lived in North Kerry for over forty years and walked this beach countless times My curiosity turned into a voyage of discovery of the voyages of *the Thetis* between 1829 and 1834.

The Irish Famine and immigration from Ireland between 1840 and 1847 have rightly received the most attention. But tens of thousands of people left Ireland in the decades before the Famine to settle in America and Canada. This is not a book about the Great Famine as the pathogen *Phytophthora Infestans* that destroyed the potato crop did not appear in Ireland until 1840 long after *the Thetis* was wrecked in 1834.

Some of the following is supported by documented facts, part is local knowledge but part of it is guesswork and imagination. Because of the absence of records or captain's logs from *the Thetis,* there are leaps of faith in this version of the story. Some questions were easily answered, but for some it was impossible to find satisfactory answers. What follows is my theory about the fate of *the Thetis*. There are obvious parallels between 2023 and the 1830s, as the authorities in Ireland and Canada were concerned about drug smuggling, organised crime, emigration, contagious diseases, quarantines, and racism.

There are many sources of information quoted here, some contemporaneous accounts, some travel books of the 1800s and even some fictional references. Robert Whyte's *Famine Ship Diary 1847* is about life on board *the Ajax* as she sailed from Dublin to Quebec City. This diary is a valuable source of information even though his name may not have been Robert Whyte, the ship may not have been called *the Ajax* and even the captain was not identified. While Whyte was a cabin passenger with relative comfort on board, his concern for the immigrants in steerage is evident throughout his diary.

In the 1994 edition of Whyte's diary, Norita Fleming's impassioned speech where she objected to the Canadian Government plans to create a theme park on Grosse Ile is included. I say this to make it clear that I am aware of the sensitivities regarding Grosse Ile among many Irish and the wider diaspora.

Brian Goggin's *Waterways and Means* has a chapter dedicated to the Shannon Estuary which he subtitles 'A 19th Century Motorway'. This book documents the efforts by determined individuals to improve safety standards along the estuary.

Paddy's Lament (1987) by Thomas Gallagher, an American writer, is another source of information about Irish immigrant ships. Gallagher's book contains graphic descriptions about life on board, particularly the lack of sanitary facilities on immigrant ships.

Martin Doyle's book written in the early 1830s is a detailed guide about emigration to Upper Canada. It contained advice to immigrants on the routes to take, prices and what to expect on board the ships crossing the Atlantic.

However, the book that made the most lasting impression and the one liberally quoted here is Edwin C. Guillet's *The Great Migration*, which was published in 1937. Guillet was a Canadian author who wrote on a variety of topics about life in Canada. *The Great Migration* is an account of emigration from Great Britain and Ireland to North America in the first half of the 19th century. *The Thetis* arriving in Quebec City from Limerick is mentioned in *The Great Migration*. But while Guillet was forensic in describing the people who emigrated he doesn't deal with the ships or the men who sailed on them in any detail.

This is an attempt at filling that gap with a mixture of cut and paste, some mathematics, a small bit of fiction, some history and unsupported opinions.

Paul O'Dowd
2024

ONE

The Shannon Estuary

The first man to sail fragile ships in the deep ocean wore armour of oak and three layers of bronze around his chest.

– Horace

In April and May of 1834, six small wooden ships left Limerick carrying 1500 Irish immigrants bound for Canada. Two of the ships, *the James* and *the Astrea* were lost at sea and 480 passengers were drowned. The other ships, *the Breeze, the Priscilla, the Albion* and *the Thetis* arrived safely in Quebec City six weeks later with 1020 passengers. The ships had sailed from Limerick due west to Loop Head at the mouth of the Shannon Estuary, before setting out across the Atlantic.

The Thetis sailed to Quebec with hundreds of immigrants and returned to Limerick with timber from the forests of eastern Canada on many occasions. Clare is on the northern side of the estuary, with counties Limerick and Kerry on the southern side.

Loop Head guards the northern entrance to the estuary with Kerry Head on the southern side. Beale Beach is located on the Kerry side, about seven kilometres from Loop Head. The beach is covered twice a day by the tide that reaches to the sand dunes. These dunes run along the entire six kilometres of the beach. The estuary here varies in width from one kilometre to over five kilometres at its widest point. The word *beal* means mouth in Irish and is generally believed to be the origin of the beach's name because it is close to the entrance to the estuary. An older Irish name for the area is *Biaile* which is said by some to have its roots in the French *Beau Lieu* or beautiful place.

On the beach, lie the remains of a ship. Timber spars protruding from the keel are all that are left. These spars were cut from softwood trees but close examination reveals hardwood dowels between the softwood spars. These dowels are in perfect condition despite being submerged for 190 years.

The wreck is covered by sea and sand most of the time but on the low spring tides, the remains of the ship are uncovered. The National Monuments Service protects the wreck prohibiting unauthorised excavation. Wrecks over 100 years old underwater, irrespective of age or location, are protected under Section 3 of the National Monuments (Amendment) Act 1987.

A short distance out from the beach is a reef known as the Beale Bar. *The Thetis* went aground on the reef as it returned from Quebec City bound for Limerick with a cargo of timber. *The Thetis* was blown on to the reef by winds from the west on 31st November 1834 – or so the story goes. But is this true? Did a storm cause the wrecking? *The Thetis* was carrying timber from Quebec in Canada to Limerick. However, *the Thetis* was also carrying bales of tobacco which in 1834 in Ireland was contraband. The ship was smuggling tobacco.

Beale Bar dries out on the lowest spring tides of the year. The rock is exposed and when this happens the reef can be seen from the beach. During other tides, the sea breaks over the rocks creating a long line of white surf. These days, Beale Bar has its own cardinal marker to warn ships coming into the estuary to stay to the north. The Shannon flows through the estuary on an outgoing tide at a rate of four to five knots and the river is tidal as far as Limerick City, 60 miles inland. In the early 19th century, sailing ships on the Estuary were dependent on

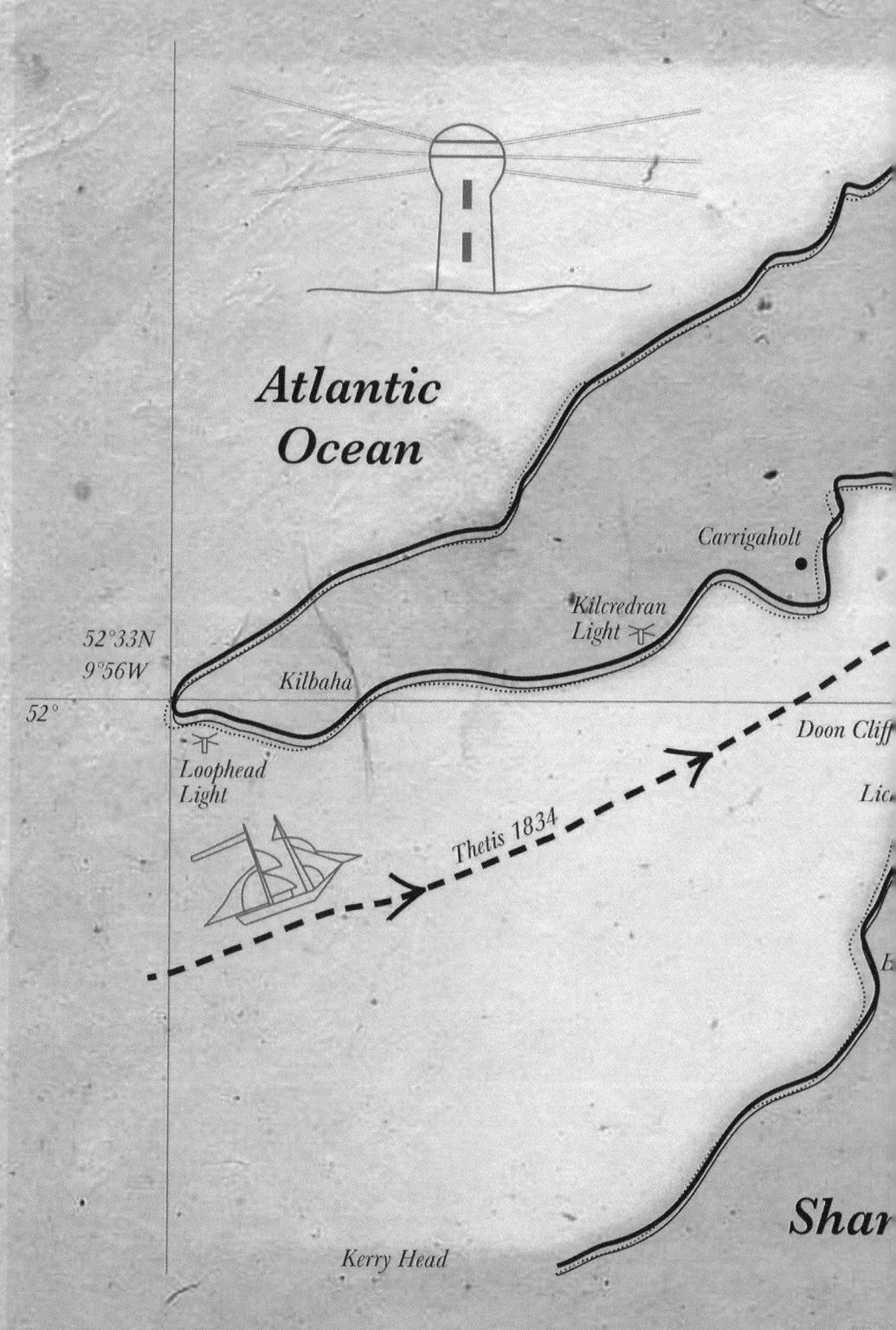

the state of the tide because they couldn't sail against the current. On arriving at Loop Head, ships had to wait for the incoming tide before entering the Estuary. In the other direction, ships leaving Limerick Port passed through a winding river to Foynes, Glin, Tarbert Island, Scattery, Kilrush and onwards to the open sea at Loop Head.

But let's start the enquiry about the last voyage with some questions. What type of boat was *the Thetis*? Who owned it? Was it seaworthy? Who was the captain? Who were the crew? How many men were on the ship?

In the 1820s, *the Thetis* along with many other ships started sailing from Limerick to Quebec. Timber was in short supply in Ireland and Great Britain, as Napoleon had imposed an embargo on the importation of timber from the Baltic States. However, timber was plentiful in Eastern Canada. A fleet of ships crossed the Atlantic from Irish and British ports bringing back large quantities of pine. In Britain, much of the timber was used to build ships which in turn were used to bring pine from Canada to build more ships.

But another cargo was being brought from Canada. Tobacco smuggling was common along the Kerry and Clare coasts in the 1800s. In 1841, for example, in a well-documented case, custom officials boarded the brig *Maria Brennan*, which was en route from Quebec to Limerick, off Scattery Island in the Shannon Estuary. They found four cwt of tobacco on board. The captain and the crew of the *Maria Brennan* were charged with smuggling, but the case was suspiciously dismissed on a technicality.

Nine sailors from *the Thetis* were reported drowned that night in November 1834. Customs officials arrested the captain and the remaining crew some days later in Tarbert about ten miles along the Estuary on the way to Limerick. They were brought to Tralee where they were tried for smuggling tobacco. But there is no record of this court case or of any subsequent penalties.

Why were the remaining crew in Tarbert three days after *the Thetis* was wrecked when nine of their crew members had drowned?

Should they not have been searching for their missing shipmates? Sometimes, when walking the beach, I hear a small voice in my mind, saying if you're so curious about what happened why don't you tell our story and unravel the mystery? And I found that when I go off track, the voice on the beach nudges me back in the right direction.

No, no not there, more to the west. Come on you know how the tides work. How can the wind blow from there? Have you checked the anchors? You're an engineer, you work it out.... I'm not going to tell you everything, where would the fun be in that?

And so on. Just hints and no answers, but never saying you're right, only a silence which I take to be a sign of approval. The voice, I hear on the beach is friendly and helpful, perhaps realising this could be the last chance to explain the voyage of *the Thetis*. I like to imagine it must be one of the crew perhaps the navigator on *the Thetis*. Perhaps, the sailor who knew how to find his way across the Atlantic to Quebec City using the position of the sun, moon and stars.

In 1834, the Age of Sail was coming to an end. Sailing ships were making way for steel-hulled steamships, even though the last sailing cargo ships were still operating well into the 20th century. They were used on voyages to the South Atlantic as they couldn't carry enough coal to reach their destination and then return from it. Other technological advancements were happening at this time, particularly in the science of navigation, more of which later.

Francis Spaight was the owner of *the Thetis*. In the years from 1820 to 1850, Spaight was a successful businessman in Limerick. The hardware company he founded, Spaights of Limerick was still operating well into the 20th century. Francis Spaight's involvement in clearing large estates of small farmers including those on his own lands, is a matter of record.

However, we don't know what he looked like. There is no portrait of him, which is unusual for a man of such importance. But there must have been a painting of Francis Spaight at one time. He was a wealthy man from an important military family and part of the landed gentry, so he must have sat for a portrait during his life.

Remains of The Thetis on Beale, as seen from another angle.

Perhaps, an unknown painting still exists hanging on a wall somewhere or lying unwanted in an auction house. After all, portraits exist of his father, William Spaight, his mother, Millicent Anne Spaight (nee Studdart) and his son James. Does this matter in the long run? Probably not, but as Francis Spaight is such an integral part of the story of *the Thetis*, it could be an additional piece of the jigsaw.

The Spaight family were not absentee landlords. They were Irish, from Munster, and were involved in all aspects of life in Limerick. James Spaight was president of the Limerick Chamber of Commerce and Deputy Lieutenant for Tipperary. He was elected unopposed as MP for Limerick City in 1858, but then lost the seat at the next general election in 1859.

We know the name of only one member of the crew who was on board *the Thetis* in November 1834. Joseph Younghusband was the captain of the ship when she went aground in Beale. He was born in 1803, but we don't know where. The name Younghusband was uncommon in Ireland, so he may not have been Irish. He might have been from the north of England because in the late 1830s, he spent time working on ships out of Liverpool. But, for the years between 1819 and 1834, Younghusband worked on ships sailing from Limerick to Canada. In 1843, he was convicted in London for

William Spaight and Millicent Anne Studdart. (Francis Spaight's Father and Mother.)

James Spaight. (Francis Spaight's son).

deliberately scuttling a ship called *the Native* which was also owned by Francis Spaight. Younghusband was transported to Norfolk Island off the coast of Australia where he died four months after arriving, aged 41. This episode is dealt with later. And that's about it, no other details about Younghusband or other members of the crew are available. Frustratingly, the names of the rest of the crew who were on board *the Thetis* in November 1834 are lost.

When *the Thetis* was wrecked on the Shannon Estuary, Joseph Younghusband was only 31 years old. Church records in Limerick reveal that a baby boy named Joseph Younghusband was baptised in St. Michael's Church, Limerick on 10th August 1832. The boy's parents were named as Joseph Younghusband and Mary O'Connor. A daughter Anne was born on November 1833. Four years later in 1837, a third child, a girl, Johanna, was born. And according to a later record, there was an unnamed fourth child born to the Younghusbands.

This is interesting, because two of the children were born three years after the wreck of *the Thetis*. Obviously, Joseph Younghusband

19

had remained in Limerick after *the Thetis* went aground in 1834, even though Francis Spaight had supposedly fired him. What happened to Younghusbands' wife and the children when he was transported to Norfolk Island? There is a record of a Mary O'Connor remarrying in Limerick some years later who may have been Joseph Younghusband's widow.

Another man, John William White was a captain on the Shannon Estuary. In the early 1830s, he skippered *the Dover Castle* steamship which made regular trips from Limerick to Kilrush carrying goods and passengers. White was not involved with *the Thetis* or any transatlantic voyages, but years later he joined Younghusband on a ship called *the Native*. White's sailing career ended when he was tried in London with Younghusband for scuttling the ship. John William White was one of 500 convicts transported on *the Anson*, on the 23rd September 1843. As mentioned previously, this will be dealt with in a later chapter..

Two men called Gorman, Timothy and Daniel, served as captains on Francis Spaight's ships. The Gormans were brothers from Kilrush in Co Clare. Daniel Gorman was highly regarded as a skipper on the immigrant ships to Quebec. He had captained *the Thetis* across the Atlantic to Quebec in 1832. Daniel Gorman had a reputation for trying to keep the numbers of passengers on board within the legal limits. He was also known to insist that ship's holds be disinfected regularly as a protection against disease. A newspaper of the time called him the "lucky captain". He died in Limerick and was buried in St. Munchin's graveyard in the city where an anchor marks his grave.

Timothy and Mary Gorman (1870).

Timothy Gorman was captain of the *Francis Spaight* in 1835 when the ship was at the centre of one of the most notorious incidents of cannibalism at sea. A young cabin boy, Patrick O'Brien was sacrificed to save the lives of his crew mates. The extraordinary career of Timothy Gorman continued until 1868 after more adventures at sea and voyages across the Atlantic. There is a detailed account of Timothy Gorman's life at sea in a documentary called *The Custom of the Sea*, as these acts of cannibalism at sea were called.

About sixteen crew members were on board *the Thetis* when the ship arrived back from Canada in November 1834. As mentioned earlier, we don't know any of their names except for the captain, Joseph Younghusband. This is strange given that nine of the crew were reported lost when the ship foundered in Beale. There should be records of an inquiry into the loss of so many seamen. But there is no evidence of an inquest into these drownings. There was only one newspaper that mentioned loss of life. The other five local newspapers simply said *the Thetis* was stuck on Beale Bar but that the cargo was saved.

In April 1834, there were 217 immigrants on board *the Thetis* when it left Ireland en route to Quebec, but the only names we know of are the six passengers who died on route. Their names were recorded when *the Thetis* arrived in Gross Ile, the quarantine island on the St. Lawrence River near Quebec City. Passengers on ships from Great Britain and Ireland had to quarantine on the island to prevent the spread of disease to the city. For reasons we will discover later, Gross Ile is a location that casts a long shadow over Irish history.

The head of the Coast Guard on the Shannon Estuary was Joseph Dexter, from Ballylongford, Co. Kerry, a village about six km along the estuary from Beale. Dexter and Lt. Triphook, who was Commander of the Revenue Cruiser, *the Hamilton*, boarded *the Thetis* a few days after she went aground in November 1834.

In 1858, a headstone was erected in a graveyard in Ballylongford commemorating Joseph Dexter's life.

> TO THE MEMORY OF
> **JOSEPH DEXTER ESQR**
> of Ahanngran
>
> Who Departed this life July the 26
> 1858 Aged 57 years
>
> Formerly of HM Royal Navy
> And Inspecting Chief Officer of
> Coast Guard
>
> **THIS STONE IS ERECTED BY HIS**
> **Affectionate Widow**

In 1819, James D'Ombrain, a former Royal Naval Officer, and a revenue official was appointed to tackle the problem of smuggling around the Irish coast. D'Ombrain was Ireland's first Coast Guard Inspector General. Originally called The Preventative Water Guard, the Coast Guard was started in England in 1809 to stop smuggling by using small boats to patrol the coast. During his travels around Ireland, D'Ombrain became aware of the extent of poverty on Ireland's West coast. He tried, on many occasions, to get his political masters in Dublin and London to deal with the dreadful living conditions of so many impoverished people that he had encountered. Sir Randolph Routh, Chairman of the Famine Commission complained to Charles Trevelyan, who was Head of Treasury, about D'Ombrain. He

Sir James D'Ombrain.

accused him of being overzealous in his representations to the Government about the lack of action during the famine. James D'Ombrain is remembered by Seamus Heaney in his poem for urging "free relief for famine victims".

Routine patrol off West Mayo;
Sighting a rowboat, heading unusually far beyond the creek,
I tacked and hailed the crew, in Gaelic.
Their stroke had clearly weakened, as they pulled to,
From guilt or bashfulness, I was conjecturing, when,
O my sweet Christ, we saw piled in the bottom of their craft
Six grown men, with gaping mouths and eyes bursting the sockets, like spring onions in drills.
Six wrecks of bone and pallid, tautened skin.
"Bia, bia, bia," in whines and snarls their desperation rose and fell, like a flock of starving gulls
We'd known about the shortage, but on board they always kept us right, with flour and beef,
So understand my feelings, and the men's, who had no mandate to relieve distress.
There was relief available, in Westport, though these poor brutes would clearly never make it.
I had to refuse food:
They cursed and howled like dogs, that had been kicked hard in the privates.
When they drove at me with their starboard oar, (risking capsize themselves)
I saw they were violent and without hope.
I hoisted and cleared off.
Less incidents the better.
Next day, like six bad smells, those living skulls, drifted through the dark of bunk and hatches
And once in port, I exorcised my ship,
reporting all to the Inspector General.
Sir James, I understand, urged free relief for famine victims,
in the Westport Sector and earned tart reprimand from good Whitehall.

Let natives prosper, by their own exertions;
Who could not swim, might go ahead and sink.
"The Coast Guard with their zeal and activity are too lavish", were the words,
I think.
– 'For the Commander of the Eliza' by Seamus Heaney

And that's about it. Those are the names of people that I can find who were involved with *the Thetis*. Or maybe I'm not looking in the right places. But I have other questions.... *the Thetis* sailed a second time to Canada in the summer of 1834, but this time there's no record of any immigrants being on board.

When they returned from this trip, why were they on the south side of the estuary? The safe channel is to the north along the Clare coast, and not along the Kerry coast. This part of the estuary is now an anchorage for large tankers going to the factories and power stations further up the river, so is generally considered safe and sheltered. After passing through the channel between Kilcredan and Lick Point, the sea calms and it is not often rough. Conditions are never rough enough to drive a ship two miles across the estuary, and certainly not by either a westerly or the prevailing southwesterlies winds. Those wind directions would push a ship like *the Thetis* safely up the estuary towards Scattery Island. On the rare occasions when it is rough, there is safe anchorage on the Clare coast in Kilbaha or Carrigaholt.

When entering the Estuary in 1834, ship captains like Younghusband, knew where they were, as accurate maritime charts of the Irish coast were available to them. In 1751, Murdoch Mackenzie, a British surveyor, was commissioned to survey the entire coast of Ireland. The survey resulted in the publication of two volumes of charts in 1774 and 1776. Mackenzie conducted a ten-year survey of the Irish coasts before his return to Great Britain's western coasts. In 1750, he wrote, 'The lives and fortunes of sea-faring persons, in great measure, depend on their charts.'

The Shannon Estuary had also been surveyed in 1779 and a detailed marine chart was available to sailors. The survey had been carried out by Joseph Huddart from Cumberland,. who was one of the most eminent engineers of his time. Huddart was familiar with the Irish coast as he had spent time as captain of a brig, selling smoked fish in Irish ports which was his family business. Huddart had another claim to fame as he was responsible for major improvements in the manufacture of rope, a critical factor in improving safety at sea. On board *the Thetis,* Younghusband must have had Huddart's charts available to him.

Joseph Huddart.

Sir Charles Trevelyan of whom it was said:
'That man is almost always on the right side in every question; and it is well that he is so, for he gives a most confounded deal of trouble when he happens to take the wrong one.'

What was happening in the 1830s?
The Dublin and Kingstown Railway which opened in 1834, was Ireland's first passenger railway. It linked Westland Row in Dublin with Kingstown Harbour (Dún Laoghaire) in County Dublin.

TWO

The Thetis

The Thetis: A female figure from Greek mythology who appears in multiple guises, including a sea nymph, a goddess of water, and one of the fifty Nereids.

As mentioned, *the Thetis* was owned by Francis Spaight, a Limerick merchant whose father and grandfather had both served in the British Army. *The Thetis* might have been Francis Spaight's first ship, but it certainly wasn't his last. During the 1830s, Spaight owned many ships that sailed from Limerick to North America to collect much needed timber from the forests of Canada. However, this changed when *the Thetis* was used to bring immigrants, or 'paupers' as they were described by Spaight, from Limerick to Lower Canada.

The Thetis was a type of sailing ship known as a brig. There are detailed records of similar ships from the same era so we know what *the Thetis* looked like. Brigs were popular in the early 19th century because of their speed and were commonly used to transport goods over long voyages. Most of the ships arriving into Quebec City from Britain and Ireland were brigs. They were more manoeuvrable than other sailing vessels such as brigatines, clippers or schooners. Brigs had also been used as prison ships, which explains why punishment cells on board ships came to be known as the brig.

A brig had two square rigged masts which were stepped in three sections. 'Square rigged' referred to the rectangle-shaped main sails on both masts. Although strictly speaking the sails were neither square nor rectangular, because the vertical sides weren't parallel and the bottom edge of the main sails were curved. Brigs had an additional large 'gaffed rigged spanker' behind the mainsail. The purpose of the spanker, triangular in shape, was to provide stability and balance to the vessel while under sail.

The size of *the Thetis* can be calculated from what remains of the ship on the beach in North Kerry. By extending the two arcs of the surviving spars forward to a point it is evident that *The Thetis* was 30m long and 8m wide with a draught of 3.25 m, which means she weighed about 220 tons. The bow is located about 7m in front of the large timber beam in the photograph of the wreck.

Some brigs had raised sections of deck to the front and back (fore and aft) of the ship. These were included to give extra headroom below deck for the captain's quarters in the rear and for the crew quarters in the front of the ship. A poop deck, French for stern, *la poupe*, is a deck that forms the roof of a cabin to the rear of the ship. The raised poop deck gave the captain and helmsman a clearer view of the way ahead. But many of these ships were flat decked throughout, which provided the crew with little or no protection from the weather or the sea.

The forecastle or focs'le was the raised area to the front of the ship. The crew's quarters were under this section. However, many of these sailing ships didn't have a focs'le. On *the Thetis*, there was almost certainly a level deck from stern to bow. Without the focs'le, the crew couldn't stand upright below deck as the headroom in their quarters on *the Thetis* was no more than 1.5m. A flat decked brig also meant there were no private cabin passengers on board *the Thetis*.

This photo reproduced on page 28 is of the brig *Ornen* from Denmark which was launched in 1842. The recorded dimensions of *the Ornen* are almost identical to *the Thetis*. The Danish ship was 31m long, 9m wide with a draught of 4.2m. It was also flat decked from aft to stern like *the Thetis*.

The Ornen, a ship of the Royal Danish Navy, 1850 (c).

The Beagle, made famous by Charles Darwin, during his voyage to the Galapagos Islands from 1831 to 1836, was a brig similar to *the Thetis*. Detail drawings of *the Beagle* from 1832 are available and these give an accurate picture of how the *Thetis* looked. *The Beagle* had two decks with living accommodation for the crew, captain and passengers on the first deck. Cargo and goods were stored in the lower deck. Admiral Sir B. J. Sullivan, who served on *the Beagle* as a lieutenant for the first two voyages (1830–1836), described the ship in a letter dated 12th December 1884 to Francis Darwin, son of the naturalist, as "very deep-waisted, that is, had high bulwarks for their size, so that a heavy sea breaking over them was the more dangerous".

A wave breaking wave over the bow of a ship is drained away by the gunnels or holes in the side bulwarks. But a second wave breaking over the bow could be calamitous if the first wave hadn't cleared the decks. Because of this flaw, the captain, Robert Fitzroy, on the later voyages of *the Beagle*, organised a full refit of the ship. There was controversy at the time, as the refit cost almost £8000, about £200,000 today. The modifications included installing a poop deck and a foc'sle. He also added a third mast, changing the status of *the Beagle* from a brig to a bark.

Robert Fitzroy.

The same sea filled our decks so deep that if another had followed it, it is not difficult to guess the result. ... At last the ports were knocked open and she again rose buoyant to the sea.
– *Charles Darwin's Beagle Diary*, Charles Darwin.

Robert Fitzroy, who was very religious, fell out with Darwin after the publication of *The Origin of the Species* in 1859. When the book was published, Fitzroy was upset about his role in the theory's development, as it proponded the view that humans evolved from less complex animals. He said that *The Origin of Species* had caused him "the acutest pain". Fitzroy is remembered here because his later work in meteorology contributed greatly to improved safety at sea. He was a pioneering meteorologist who made accurate daily weather predictions which he called by the new name of "weather forecasts". In 1854, he established the British Met Office.

The Thetis was a cargo ship which was used to bring timber across the Atlantic, so little attention was paid to the comfort of either the crew or even the captain. Steerage passengers could expect no regard for their comfort. At the time, none of these ships had bulkheads installed. Bulkheads were partitions that divided the hull horizontally so if a leak occurred in one section, it could be contained to that section. Without bulkheads, the entire hold and steerage area flooded when the hull was breached. This absence of bulkheads almost certainly contributed to the demise of *the Thetis*.

There were two lower decks on *the Thetis*. The captain's quarters, at the back of the boat, included a storeroom, a table for eating and for laying out navigation charts. The front of the ship, the bow, was reserved for the crew. Their sleeping quarters, with hammocks and mess table were in this section.

Spare masts, yards, and various other gear were lashed to the deck. More space was needed to store equipment such as spare sail canvas. Many different types of ropes were needed as replacements. All this spare canvas and ropes took up space in the lower decks.

The storage of fresh water was very critical. Water had to be securely stored in barrels and kept under the control of the captain when large numbers of passengers were on board. Most journeys across the Atlantic took about 40 days, but occasionally ships spent twice that time at sea. Water for drinking and cooking for large numbers of people for up to 90 days at sea was vital for survival. There needed to be 10,000 gallons of water on board *the Thetis* to cater properly for the needs of the 240 people. Records of other ships of the period would suggest that there was nothing like that amount of water on board *the Thetis*. A barrel contained about 50 gallons so at least 200 barrels should have been stored for the voyage. But there wasn't room for that many barrels on board *the Thetis*. Many ships had less than half the required amount of fresh water on board. Keeping the water fit for drinking was a constant problem on these voyages. Various ways were used to keep the water suitable for drinking. Adding vinegar to the water was one method, apparently with limited success. Water had to be strictly rationed by the captain, especially if the voyage across the Atlantic took longer than anticipated.

Further storage on board the ship was reserved for coal, which was needed for cooking. Many tons of coal were required to cater for cooking for the 240 people on board.

A large quantity of rope was needed, even on small ships like *the Thetis*. Ropes were used for a variety of tasks on board. They were needed to control the sails, to hold the 30m masts in place, to moor the ship and to raise the anchor, to mention just some of the uses. The ship was fully rigged when Spaight bought *the Thetis*, but this rigging wouldn't have lasted as ropes wear out with use and had to be replaced frequently. The standard length of a single rope on a British Naval ship was 300m. A sailing ship such as *HMS Victory* required 50km of rope.

The Thetis was hardly a first-rate ship, but many kilometres of rope were required on even the most basic of sailing vessels. There had to have been at least 10 kilometres of rope on board *the Thetis*. The largest rope maker in these islands was the Chatham Docks Ropery in the southeast of England, which was built in 1728. It was extended in 1812 when hemp houses, yarn houses and a double rope house were built. A hatchelling house was also constructed; the first step in rope-making where hemp was combined before spinning. The ropewalk was 346 metres long, and at the time of construction was the longest brick building in Europe, and had room to make a 300 metre rope by hand.

Where Spaight bought replacement sails and ropes is lost to history, but there are clues left behind. Some ropes, known as sheets, had to be continuous to run smoothly through the wooden pulley blocks and up the mast. To manufacture the longest lengths of rope of up to 300m, long, straight roads, some covered, but many uncovered, were needed. Streets called 'ropewalk' are still found in many places in Ireland, such as Sligo, Skerries and Ringsend in Dublin. Near Arthur's Quay in Limerick, what is now Grattan Street was originally called Ropewalk. In Ireland, the largest ropemaker by far at the time was the Belfast Rope Company.

The Belfast Ropeworks Company is known over the world, not only as being the largest in existence, but also for the variety and excellence of its products. It now covers an extent of thirty-four acres, and gives employment to many thousands. Until the 1740, all cordage was imported from England, then ropemaking place was commenced by John McCracken in 1758. Thomas Ekenhead was for many years the principal ropemaker in Belfast at 56 Anne Street.
– *The Story of Belfast and It's Surroundings,* Mary Lowry (1913).

Making sails for sailing ships was a very precise manufacturing process. Many ships had twenty sails of different dimensions which were cut using exact templates. The larger ships of the British Royal Navy allocated a large space below decks where their own sailmakers cut and sewed the sails from large rolls of cotton.

Napoleon's embargo on goods being exported to Britain and Ireland, affected the cost of providing sails. Margaret Robinson's

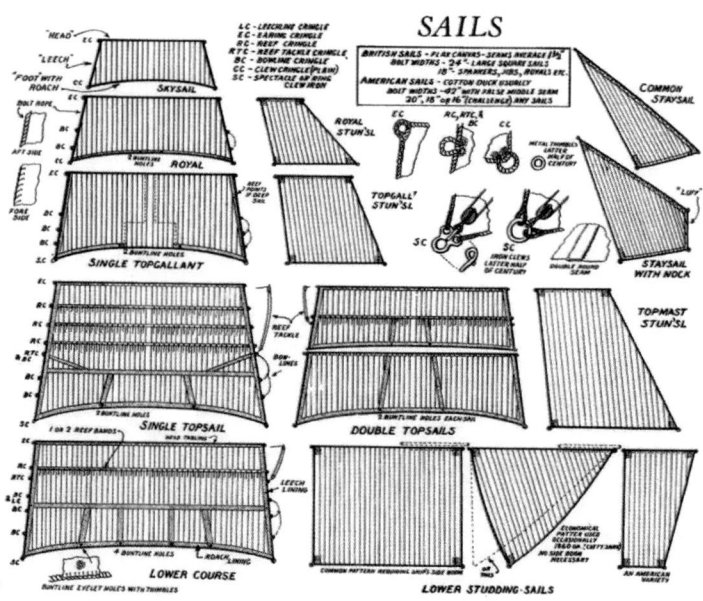

This drawing is an example of the complexity of sail manufacturing.

book, *Lancaster's Sail-Cloth Trade in the Eighteenth Century*, explains the increase in price of the manufacture of sailcloth in the 19th century.

Raw material prices had risen sharply during the Napoleonic blockade and no.1 canvas had gone up from 13.5d. a yard in 1785 to 15.5d. in 1792, to 20d. by 1807. The two-masted brig Flora had a mainsail of 146 yards of no.1, a square mainsail, 165 yards of no.3 and two topsails of 150 yards of no.2. – Lancaster's Sailcloth Trade in the 18th Century, Margaret Robinson.

A full set of sails for a two masted brig like *the Thetis* cost £100 for the material in 1820. Adding in labour and inflation costs would mean at today's prices the sails cost about £5,000. This seems remarkably cheap given that a single racing jib sail for a yacht can cost €4,000. Presumably, Francis Spaight would not agree.

There were three main elements making up the construction of *the Thetis,* cotton, timber, and hemp. Cotton was used for the sails and hemp for the manufacture of rope. The timber, for the hull, decking and masts, came from the Baltic States prior to Napoleon's embargo.

The timber hull of *the Thetis* kept the ship afloat, the sails propelled the boat forward and the ropes controlled the sails. But what stopped the ship? Taking down or furling the sails reduces the speed of the ship but currents, tides or wind continues pushing the boat forward or backwards. The only way to bring a sailing boat to a complete stop (apart from running aground) is to throw a heavy weight overboard with a rope or chain attached to the ship. And it has been that way for millennia. But unless the weight is equal to the weight of the ship something else must happen. The anchor must bed in the bottom of the sea. This may seem obvious, but anchor science and developments have been ongoing for centuries. The 19th century saw major advances in anchor technology.

The anchor was one of the most important pieces of equipment on board sailing ships. Most ships of the 1800s carried a variety of anchors. These ships couldn't operate without anchors, because the anchor was the brake to stop the ship. The anchor had to be properly sized for the ship and its cargo. And yet sometimes it wasn't… There was 100 m of chain or rope attached to the anchor for mooring the ship even in relatively deep waters.

An example of an Admiralty anchor, in Foynes, Co Limerick.

In the early 1830s, on ships like *the Thetis,* the anchor most in use was The Admiralty Old Pattern Long Shanked anchor. The photograph above is of an Admiralty anchor with an iron stock, which became the most popular anchor in use in the mid 1800s. It forms part of a display in Foynes, on the Shannon Estuary. Some of the anchors had a timber stock instead of the iron stock. The Admiralty anchor has arms at right angles to the flukes at the end. When the anchor was dropped, one of the arms kept the flukes in the correct position which ensured that the anchor bedded firmly in the sea bottom. The anchor on *the Thetis* was probably an Admiralty anchor and apart from a timber stock was roughly the same size as the anchor in Foynes. These anchors are displayed as monuments in many Irish ports, including Crosshaven, Foynes, Limerick, Fenit, Salthill and Dunmore East to mention just a few.

When thus far advanced, a seaman rebuked the deficiency, by asking if a ship, completely rigged, was to remain an inert body. Of what use, said he, are these masts, and stays, and braces; these blocks, and sails, and anchors? .... Hence was perceived the necessity of an union between the naval arts and the purposes to which they are applied.
– The Elements and Practice of Rigging And Seamanship, David Steel.

The size of the anchor for a particular size of a ship was determined by various guides, one of which was William Sutherland's *Britain Glory or Ship-Building Unvail'ed* (1717). His formula for the length of the anchor shank was 2/5ths the beam of the ship at its maximum width. *The Thetis* was 8m wide so the main anchor shank should have been 3.2m long. This anchor weighed about 9-10cwt or 450kg. This was probably the type of anchor on board when the ship was first brought to Limerick in 1819.

The first major change in anchor design was the Hawkins' patent of the 'Stockless' anchor in 1821. The picture below shows a monument to the seamen lost on the Shannon Estuary. It is located at Mount Kennet in Limerick on the bank of the river.

The ships, ropes and canvas sails may be gone, but the cast iron anchors are erected in harbours around Ireland, as a last reminder of the Age of Sail.

The anchor in the monument is a stockless anchor. Stockless anchors consist of a set of flukes (the blades at the bottom of the anchor) connected by a pivot or ball and

A stockless anchor, located in Mount Kennet, Co. Limerick.

socket joined to a shank. The stockless anchor became popular very quickly as it was easier to store on the ship after lifting it from the seabed. The flukes rested against the hull instead of having to be brought on deck.

The Thetis should have had at least four anchors on board. Two main or bower anchors, a stream anchor, and a kedge anchor. Hauling in a nine cwt ship's anchor and a length of chain or rope often 100m long required hard physical work. In the early to mid-nineteenth century, on larger ships, anchors were hauled on board with a windlass, which is a horizontal drum with a ratchet attached. Sailors achieved leverage by pushing long capstan bars to turn the windlass and raise the anchor. Many of the smaller brigs like *the Thetis* had no mechanical aids and relied entirely on manpower to retrieve the anchor. And therein lies a problem.

In an emergency, without the right size anchor for a fully loaded ship and enough crew to retrieve the anchor, the result could be catastrophic for the ship, as we shall see later. *The Beagle* on its voyages to the Galapagos had an extraordinary number of anchors on board, eleven in total and a scarcely believable two km of chain. There were five 14 cwt bowers or main anchors, two seven cwt stream anchors for rivers, and four three cwt kedging anchors. The fate of *the Thetis* in November 1834 was almost certainly linked with the type of anchors on board or more to the point, with the absence of the correct anchors on the ship.

Ships carried a rowing boat for going ashore when at anchor. This boat was stored on the deck between the two masts. Access hatches took up more space on the deck, along with rudimentary toilets and cooking fires. All of which left little recreational space on the main deck for the 200 passengers. The lower deck or steerage was where the passengers were accommodated. The area below decks was very cramped without any daylight or very often any form of ventilation. In bad weather, the hatches were battened down keeping the passengers locked in without access to fresh air, toilets, or cooking facilities.

The rear section of the main deck, about six metres in length, was where the wheel and the compass were located. This section was out of bounds for everybody except the crew. The bow area narrowed to a point further restricting space on board. With 217 passengers and eighteen crew on board *the Thetis*, this was a very crowded deck.

The Thetis had no suitable passenger accommodation. It was built to transport goods across the Atlantic, not people. *The Thetis* was neither built nor maintained to Navy standards of 'ship shape and Bristol fashion'. *The Thetis* was the sail equivalent of a tramp steamer. Swabbed decks, polished brasses and gleaming white sails were not high on Spaight's priority list. But *the Thetis* had been across the Atlantic several times so the ship had to be serviceable, if in need of some tender loving care.

This photograph of the brig *USS Niagara* was taken in 1913. American forces had deliberately sunk *the Niagara* during the War of 1812 with the British. The ship was later restored, or, perhaps more correctly, rebuilt. But, if someone had a camera in Limerick in

An image of the USS Niagara (Brig) 1913.

1834, unlikely as the camera wasn't invented until 1839, then this is what *the Thetis* would have looked like apart from the cannons. And of course, then, as with *the Thetis,* there is a steamboat lurking in the background ready to take over.

A disadvantage of brigs was that they couldn't sail as close to the wind as other sailing vessels. This problem often determined the course to take across the Atlantic. However, on the plus side a brig could be manoeuvred under sail in restricted spaces. They were suited for sailing to small harbours where rowing boats couldn't assist in mooring the ships.

The timber hull and decks were very basic and were built using centurys-old techniques. Above deck, things were very different and much more sophisticated. Brigs had a complicated arrangement of sail management, known as 'running rigging'. This was used to control the main sail, the topsail, the top gallant and the royal sail on the main mast. On the foremast, there was a similar arrangement for the foresails. This drawing of *the Marie Sophie* shows how intricate the sails, ropes, masts and spars could be. Although, it is probable that *the Thetis* had only three or four sets of sails on each mast and not five as shown on *the Marie Sophie.*

Two Years Before the Mast is a memoir by the American author Richard Henry Dana Jr., that was published in 1840. Dana wrote his account about a two-year voyage from Boston to California, which started in 1834. He was aboard a brig called *the Pilgrim* for part of the voyage and he gave a vivid account of life at sea.

The Pilgrim was a smaller brig than *the Thetis* at 27m in length. The numbers of crew on board *the Thetis* were roughly similar to the numbers on *the Pilgrim*. As described by Richard Dana the ship's complement included three officers Captain, First Mate, Second Mate and eight common sailors. The Second Mate commanded the starboard watch. There were four specialist crew members who were not part of any watch, the steward, cook, carpenter and sailmaker. *The Thetis* had a crew of at least sixteen, and perhaps eighteen at times. Usually, there were two apprentices on board these ships who were about fourteen years old. The accounts of the last voyage suggest there were sixteen crew members on board when *the Thetis* went aground.

The captain of *the Pilgrim* gave a warning to the crew after setting out.

Now my men, we have begun a long voyage. If we get along well together, we shall have a comfortable time; if we don't we shall have hell afloat.
– Captain Frank Thompson.

Richard Dana writing about life on board another brig called *the Alert* mentioned a problem that affected all brigs of this design.

There was one difficulty, however, which nothing we could do could remedy; and that was the leaking of the forecastle which made it very uncomfortable in bad weather and rendered half of the berths untenantless. The tightest ships in a long voyage from the constant strain on the bowsprit, will leak more or less around the heel of the bowsprit and the bitts which will come down into the forecastle.
– *Two Years Before the Mast*, Richard Dana.

Returning to a wet bunk after hours on watch, is a miserable experience and can happen on all sailing ships even to this day. A change of clothing and a dry bed is needed to recover after a long storm at sea. But, there is no doubt that *the Thetis* leaked continuously in the bow making the passenger accommodation very wet. It was also at times too hot, then very cold, but always cramped, dark, and smelly.

Another brig was involved in a famous maritime incident. The brig *Dei Gratia* was east of the Azores on 5th December 1872, when crew members spotted a ship. The vessel was the brigantine *Mary Celeste*. When the crew boarded *the Mary Celeste* there was nobody on board. The disappearance of the crew and passengers remains a mystery to this day.

Brigs also featured in many works of fiction such as J.M.Barrie's *the Jolly Roger* in Peter Pan, Patrick O'Brian's *HMS Sophie* in *Master*

The New Lady Washington.

and Commander and Jules Verne's *Speedy* in *The Mysterious Island.* The Horatio Hornblower series of books has also featured brigs. There are brigs in films; *Star Trek* (*Lady Washington*), *Pirates of the Caribbean* (*Interceptor*), and computer game 'Assassin's Creed' (*Jackdaw*).

This was the beginning of the end of the Age of Sail, and steam ships were starting to take over. The first steamship purpose-built for regularly scheduled trans-Atlantic crossings was the British side-wheel paddle steamer SS *Great Western*, built by Isambard Kingdom Brunel in 1838.

In Ireland, Charles Wye Williams founded The City of Dublin Steam Packet Company commonly referred to as the CoDSPCO.

In 1830, this company had twelve ships bringing goods and people from Dublin to Liverpool. In 1832, they also provided a ferry service from Limerick to Kilrush. Originally the service was by paddle steamer but in the 1830s, *the Clarence* and *the Garryowen* steamers took over. *The Garryowen* was a substantial craft, 38m long with 90 hp engines and cost £16,000. It was said to be the largest iron ship in the world and spent twenty-five years on the lower Shannon. Trips to the seaside in Kilkee were arranged when the passengers arrived in Kilrush. Pony and traps brought holiday makers to the beaches in Kilkee for the day. When *the Thetis* made her last journey down the Shannon in 1834, she must have met the steamships *the Clarence* and *the Garryowen* on their regular trips from Limerick to Kilrush. *The Thetis* was to meet *the Garryowen* once more in December 1834 when she was marooned on Beale Bar.

This was a time when travel by sea was being influenced by science and technology. Joseph Younghusband and his crew had to be competent sailors to cross the Atlantic from Limerick to Quebec City on these open decked boats. The skill required to navigate the Atlantic in the 1800s demanded a degree of literacy and numeracy. Many of these sailors were apprenticed at fourteen years of age or younger, so they had a very basic education. Yet they could carry out complex calculations and chart work to navigate across the Atlantic. How they managed to achieve this with over 200 nervous and frightened men, women and children is hard to comprehend. But the crews were not always so highly regarded, as this extract from *The Great Migration* describes:

In some vessels, there appears to have been no discipline among either passengers or crew. "We don't wish you to come with such company as we did," wrote Stephen Turner in 1828. "From the captain to the lowest sailor they were abominable wicked; and there was no order, but swearing, cursing and drinking." Similarly, Francis Thomas who crossed on the Hebe in 1833, observed in his diary that the crew were "a set of the most vile abandoned wretches I ever met with."

Perhaps Francis Thomas and Stephen Turner in 1828, might have remembered that their lives depended on these "vile abandoned wretches". Men who often had to climb 30m in the middle of night to trim the sails in the horrendous storms can hardly be dismissed as having no discipline. On land, these gentlemen were used to having the lower orders bowing and scrapping before them. On board ship, however, the order was reversed. The crewmen issued the orders and decided when passengers of all ranks were to eat, sleep and exercise. Perhaps it is no surprise that the gentlemen were resentful at being ordered about by these 'wretches'.

The Limerick ships sailed to Quebec City in the summer because the St. Lawrence River freezes over in the winter and becomes unnavigable. So, there must have been times on board when the weather was warm and the seas were calm. With the wind on the beam, the sails up and *the Thetis* sailing at eight to ten knots, parts of the journey might have been enjoyable. Occasionally, mothers were allowed bring their children on deck at night to see the stars and the rising moon.

Was everybody full of hope and excitement and looking forward to a better life? At times, there were songs, dancing and laughter. In 1833, the people on board had not endured a collapse in the potato harvest. They were better physically equipped to endure the cramped conditions on board than the later famine immigrants. But the living quarters were dark, stinking, overcrowded and unsanitary. Privacy was completely out of the question. And there was the hidden danger of infectious diseases, mainly cholera, but also typhoid and, of course, seasickness.

Seasickness affects almost everybody on board ships to a greater or lesser degree. Experienced sailors will shrug it off within hours as soon as they get their 'sea legs', but some passengers on *the Thetis* might have been ill for days. Keeping your eyes on the horizon helps restore the balance of the inner ear, alleviating the effect of seasickness. However, as the passengers on *the Thetis* were confined below decks in steerage this was not an option.

Their land-based sense of balance failed them; they watched as one by one those around them gave up the fight and ran to retch, until the power of suggestion almost equaled the power of the ship to dance with the sea. Those young or agile enough climbed through the hatchway to the main deck to vomit over the side; the very young and very old became sick where they were–in their bunks or wherever on deck they happened to be.
– Paddy's Lament, Ireland 1846-1847: Prelude to Hatred, Thomas Gallagher.

Seasickness is sometimes referred to as a contagious illness, but, in large groups of passengers, it is perhaps better described as a chain reaction. One person being ill causes a reaction in others and so it goes. The unpleasant smell acts as a trigger causing nausea to the others on board. It is no wonder intending passengers were advised by family members who had gone to America before them to book the upper berths if possible. Sailors, such as Nelson, Darwin, and Columbus, suffered from seasickness. In Ancient Rome, Horace wrote of seasickness as a social leveller, affecting both wealthy and poor.

"He is as much surfeited in a hired boat, as the rich man is, whom his own galley conveys."

Seasickness eventually subsides for most people, and then a kind of routine is established on board. Boredom on the other hand persisted for everybody on the month-long journey across the Atlantic. How did they pass the time for forty long days crossing the Atlantic? They did very little, besides sleeping and sitting in groups on deck playing cards or pitch and toss. Some ships had religious ceremonies on Sundays, often for both Catholics and Protestants. At times, in calm weather, there was music and dancing on board, most of which were Irish jigs and reels but other types of music which were popular then might have been heard on board.

Thomas Moore's *Melodies*, published annually between 1817 and 1834, was widely available in Ireland so perhaps his poem/song *The Coulin* might have been heard on board *the Thetis*. The first verse shows that Moore was well aware that large numbers of his fellow Irishmen and women were leaving Ireland.

Sir Thomas Moore (1779–1852)

Though the last glimpse of Erin with sorrow I see,
Yet wherever thou art shall seem Erin to me;
In exile thy bosom shall still be my home,
And thine eyes make my climate wherever we roam.
To the gloom of some desert or cold rocky shore,
Where the eye of the stranger can haunt us no more,
I will fly with my Coulin, and think the rough wind
Less rude than the foes we leave frowning behind.

– 'The Coulin' (My Fair-Haired Lady).

For the immigrants, trying to prepare food on board ships such as these as they crossed the Atlantic was very difficult. The Captain and the First Mate on *the Thetis* ate together in the captain's quarter directly beneath the poop deck area, which was standard procedure in these sailing ships. There was a coal burning stove located in the captain's quarters which was used to cook his food. There was a designated cook who cooked in the bow of the ship for the crew on another coal stove.

The passengers in steerage on *the Thetis* had no such luxuries. They were expected to bring food for themselves and their families for a seven-week voyage. They were given limited time each day to cook their food in coal burning braziers on the open deck.

Cooking on a modern sailing yacht while under sail is an acquired skill that even some of the most experienced sailors find difficult. These days, gas cookers are used on sailing boats. Gimbels which move with the motion of the boat keep the cooker and pots level. In the 1830s, pots and stoves had to be kept clamped in place over coal fires. The quality of food consumed by both passengers and crew varied considerably. It was critical that the crew were provided with basic fare, at the very least. The crew had to be kept fit and well, as everybody's lives depended on their ability to carry out the heavy physical work required on these ships.

Undercooked and rancid meat meant that gastroenteritis was a common problem for passengers on board these ships, which occasionally caused death. In rough weather, nobody was allowed on deck and the hatches to steerage were battened down, sometimes for days on end. Keeping meagre provisions edible in damp conditions was a constant problem.

Terror-stricken as they were in the black dungeon that had become their prison at sea, weak as they were with nothing warm to eat and only contaminated water to drink, unable as they were to use the privies on deck, they would remember for as long as they lived the horror, misery, and elemental fear of being locked like animals in steerage during a storm, when insanity was the only escape and sanity an almost unendurable burden.
– *Paddy's Lament: Ireland 1846-1847 Prelude to Hatred,* Thomas Gallagher.

For captains like Younghusband, not knowing exactly how long the journey might last was a problem. If the journey took longer than planned, people would run out of food before reaching Quebec City. So, what then? Hard decisions had to be made and self-preservation became all important to both crew and passengers. The captain and crew had to protect their food supply or else everybody's lives were put at risk. If the situation became too desperate, there were 217 hungry people on board in steerage who could easily overpower the crew.

There was a crew of about sixteen men on board *the Thetis*. They operated a system of watches, with half the sailors on deck for four hours and the others below deck resting. A 'dog watch' of two hours was used to make sure the same men weren't always on duty at the same time, every day or night. This watch system kept the changeover times moving forward day by day. In an emergency, of course, the well-known call of 'all hands on deck' was used.

Brian Murphy and Toulha Vlahou's *Adrift* is the story of *the John Rutledge's* voyage from Liverpool to New York, with 124 passengers in the winter of 1856. The account of the conditions faced by passengers in steerage can't be bettered.

The air in steerage was weirdly stagnant even as the winds howled just above. The stench quickly grew over powering. The unmistakeable acid sweet smell of vomit infused every corner. If you couldn't get to the deck the only place to retch was in your berth or on the floor.

The John Rutledge hit an iceberg on February 19, 1856, in foggy weather on the Grand Banks off the Gulf of St Lawrence and sank. Only one man, Nye, survived.

The Brig 'Martha' passing the Fastnet Rock. Painting by Joseph Heard (1799-1859).

THREE
Francis Spaight (1790-1861)

*Oh, I'm sailing away, my own true love
I'm sailing away in the morning
Is there something I can send you from across the sea?
From the place where I'll be landing?*

– Bob Dylan

Thomond Bridge and harbour, c.1826. Painting by William Stokes.

We know that Francis Spaight was the owner of *the Thetis*. But who was he? It is thought he was born in Kilrush in 1790. He died in 1861, in Limerick. His father, William Spaight, had fought with the British in the American War of Independence. His grandfather had also served in the British Army. Francis Spaight decided against a military career, preferring to go into business. In 1798, his father died when Francis was only ten years old. Perhaps, financial pressures on the family fortunes after his father's untimely death made Spaight's decision for him. Francis Spaight became involved in the import and export business.

Francis Spaight was only twenty years of age when the *Clare Journal* reported on 24 Sep 1810 that "Messrs. Paterson and Spaight of the port of Kilrush are appointed agents for supplying his Majesty's ships and vessels in the Shannon with rum."

When he was twenty-two, Francis Spaight married Agnes Patterson, the daughter of James Patterson, a naval officer. Francis and Agnes had four surviving children, George, Amelia, Harriette and James. In 1822, Francis Spaight, together with his father-in-law, James Patterson, established a shipping company.

He became the foremost ship owner in Limerick, sailing the Atlantic to Quebec City. Francis's son James Spaight became High Sheriff of Limerick City in 1853. Later, in 1856, James was Mayor of Limerick, which shows how important the Spaight family had become.

Charles Dickens might have written some sections of *Reminiscences of Old Limerick (1939)* which contains an evocative description of an exchange between Francis Spaight and Sam Evans. Sam Evans was the bookkeeper for the Spaight company and was a central figure in the keeping of records. Shades of Dickens' characters Bob Crachett or Uriah Heep perhaps which might also be just as fictional?

Francis Spaight the founder was an energetic enterprising young man with a somewhat fiery temper. The only occupant of his private office was Sam Evans his confidential clerk a solemn old Quaker with a long mosaic beard

who sat at his desk poring over his books and taking no notice of his employer's tantrums. Things did not always work smoothly and Frank Spaight's temper was often severely taxed but when more than usually exasperated he would hurry into his private office, he strutted up and down swearing like a trooper. After a while Sam Evans would slowly turn around from his desk and say "Friend Francis what if thee was called to meet thy maker tonight? I'll bet you a hundred pounds I won't be" was the hasty report. He would have his won his bet for years after he bought the beautiful estate of Derry Castle on the shores of Lough Derg and spent his old age amid its scenic beauties.
– *Reminiscences of Old Limerick (1939),* Ernest Henry Bennis.

Spaight owned many ships over a twenty-year period, not all of which were used on transatlantic journeys. His ships transported goods from Limerick to London, Bristol and Liverpool. Many of Frank Spaight's ships were involved in major disasters. He owned two ships called Thetis, both of which had sorry ends. He named two of his ships after himself. The first *Francis Spaight* was involved in a notorious incident of cannibalism amongst the crew.

During the early 1800s, up to twenty brigs similar to *the Thetis* sailed from Limerick to Quebec City most of them making two trips a year. The return trips carried timber from Upper Canada. As mentioned, in 1834, *the Harvey, the Martha, the Breeze, the Priscilla, the Albion* as well as *the Thetis* were some of the ships that made the crossing from Limerick, bringing 1500 immigrants across the Atlantic to Quebec

To try and mitigate the effect of the tides, Spaight and other Limerick merchants built a floating dock near Mount Kennet in the city. A monument was erected in 1853 in Limerick commemorating the dock's opening. The names of Francis Spaight, W.H. Hall Mayor, James Spaight, High Sheriff, and John Long, Engineer are engraved on it. The monument lies inside the security barriers of Limerick Port, completely forgotten and out of sight. It's a large ugly stone plinth, over 4m high, with little or no artistic merit, but in the right setting it could serve as a reminder of the mass immigration from Limerick in the 1800s. As the port facilities in Limerick were limited by the tide, the floating dock gave Spaight an advantage over other ships. With a floating dock, ships could come and go as they pleased.

Spaight was involved in the immigrant trade before he became a landlord. In the 1830s, landowners were clearing the lands of tenants who could no longer afford the rents. Between 1815 and 1835, there was a major collapse in the price of agricultural products. Beef had fallen from 6d to as low as 3d a pound from 1820 to 1830. The price of wheat in Limerick had dropped from 20d a stone to 13d a stone between 1821 and 1834.

Thousands of people from small land holdings left for the US or Lower Canada. They were expected to pay for their passage and provide their own food on the forty-day journey. This exodus began in earnest from Limerick in 1827 when over 20,000 Irish made the journey. The annual emigration increased to 65,000 in 1832. The numbers of people leaving Ireland from 1840 to 1847 exceeded this figure through the worst days of the Famine. But as Roy Foster points out in his *Modern Ireland 1600-1972*, there were various reasons why so many Irish were emigrating in the years before The Famine. The population of Ireland in the 1830s is generally agreed to have been about 8 million, so a combination of lack of resources and opportunities led people to look elsewhere.

To conceive of Irish Rural society in the early nineteenth century as made up of levels of society representing an agriculture that that was responding fast but unevenly to altering market forces of 'landlord and tenant' creates a dangerously simple impression. It was a complex layered structure embracing many levels of society representing an agriculture that that was responding fast but unevenly to altering markets.
– Modern Ireland 1600-1972, Roy Foster.

It has been said that Francis Spaight was motivated by the desire to help, perhaps realising that that there was no future for the thousands of people trying to eke out an existence on small holdings. There might be some truth in this argument during the famine years, but it is doubtful whether in the 1830s he was motivated by anything other than self-interest. In contrast, his son, James, showed more concern for his fellow Irishmen in his political career. But on the other hand, Frank Spaight and the other shipowners provided a safety valve that undoubtedly saved many immigrants from the devastating consequences of the Great Famine.

This may be the earliest photograph of a sailing ship in Limerick port. The photograph was taken in the mid-1840s at what was called George Street and is now O'Connell Street. The ship in the background is a brigantine and slightly smaller than *the Thetis* but it is a good indication of the type of vessel used to bring thousands of Irish across the Atlantic. The ship is shown grounded on the mud at Arthur's Quay, in an area which has since been reclaimed.

Arthur's Quay, Limerick pictured sometime in the 1840s.

One of the crew is up on the mast furling a sail and waiting for the incoming tide. The ship is high out of the water so perhaps maintenance work is being carried out on the hull. The buildings in the foreground, Hoggs Hardware on the left, for example, are long gone, but some of the houses on the far bank of the river are still in existence.

Francis Spaight became so successful in running his shipping business, that he was able to buy Derry Castle and Burgess estate in County Tipperary, a large holding of 3000 acres on the shores of Lough Derg.

The Derry Castle property.

THE DERRY CASTLE PROPERTY, *which, for its splendour and renown, stands high amongst the most favoured throughout Ireland. This circumstance is not a little refreshing, inasmuch as the writer is relieved from an attempt to do it adequate justice, and to content himself with a mere outline.*

It may be well, first, to observe that, fortunately, the Estate is free from that fearful pest to agricultural improvement and the yeomen's comfort – the middle men. All are yearly tenants; the tithe is commuted; and it is a fact of no small importance to know that the use of spirituous liquors is unknown throughout this vast district; the necessary consequence is a total absence of

  POLITICAL DIFFERENCES, OR DISTURBANCES of any kind. Having thus cleared the ground of the great difficulty that has but too frequently prevailed in the minds of THE TIMID ENGLISH CAPITALIST,
– Dublin Evening Mail, 7 August 1840.

This newspaper notice, offering the Derrycastle and lands for sale with the capital letters as printed at the time, was placed by the auctioneers. This is perhaps the most politically incorrect property "for sale" notice ever published. The "great difficulty" refers to the middlemen who had sublet the land to hundreds of small tenant farmers.

The Derry Castle and Burgess estate, county of Tipperary, was knocked down to Francis Spaight, Esq., of Limerick, for £39,500 at the chambers of Master Goold, yesterday. The highest bona fide offer for this property at the sale last May, was £37,500, and it was then bought in at £38,000. The estate now sold to Mr. Spaight comsprises [sic] 3,000 Irish acres of land, with mansion-house and offices, on the most picturesque and frequented part of the Upper Shannon, near Killaloe.
– Statesman and Dublin Christian Record
16th August 1844 and The Cork Examiner 21st Aug 1844.

Francis Spaight lived on the estate in Tipperary after retiring. His son James continued to run Spaights of Limerick and the company was still operating well into the 20th century. When Derrycastle was owned by Francis Spaight, the buildings were valued by Griffith's Valuation at £71. The *Nenagh Guardian* of 15th April 1872 reported a bad fire at Derry Castle. Perhaps those portraits of Francis Spaight were lost in the fire. In a series of articles, 'The 12 Historical Jewels of Limerick', Rachel Kealy, a Limerick journalist notes that: "The Spaights lived complicated and sometimes contradictory lives. They were industrious and hard working while also enjoying social distinction… They were tough task masters as well as active philanthropists. They were proud of their British establishment roots but made Limerick their home."

In 1851, Francis Spaight founded a school near his Derrycastle Estate called the Derrycastle Model Farm School. The school continued for a hundred and seventeen years until 1968 when it merged with Ballina National School. The ships of Limerick port flew their flags at half-mast when Francis Spaight died in 1861.

Death of Francis Spaight, Esq., J.P.

– We deeply regret to have to record the death, on Saturday morning, at Derry Castle, of Francis Spaight, Esq., J.P., President of the Limerick Chamber of Commerce, and representative in the Town Council of the Shannon Ward. – Limerick Chronicle

But what flags were flown at half-mast by the Limerick ships when Francis Spaight died? The Irish tricolour was not used in the 1830s. The maritime flags or ensigns in use at the time are pictured above with the harp representing Ireland common to all. The Union Jacks are different as St. Patrick's saltire is missing from the flag with the red background. There are clues left behind in the paintings of the period. Francis Hustwick (1797–1895) was a British marine artist who was originally from Hull. He was commissioned by the McCorkell family to paint pictures of their ships. The McCorkell Line was a shipping line founded by Wm. McCorkell & Co. Ltd. in 1778. Their ships carried passengers from Ireland, Scotland and England to America during the Famine years. One of their ships was *the Mohongo*.

The Mohongo, by Francis Hustwick.

In the early 1850s, *the Mohongo* sailed across the Atlantic from Derry on over 100 occasions bringing thousands of Irish immigrants to America and Canada. In the picture above, *the Mohongo* is flying a variety of flags. The flag at the stern is the plain 'Red Ensign' or 'Red Duster' and not the one 'defaced' by the Irish harp. The signal flags that can be identified are the red and white square flag to indicate that there is a pilot on board. The blue flag with the white square is to indicate that the ship is leaving the port. Presumably, the flags at the top of the masts are company flags.

When *the Thetis* arrived in Quebec, Younghusband would have the Red Ensign, defaced or not flying from the back stay. Many European countries were still not allowed access to Lower Canada so identifying ships origins was important to the port authorities in Quebec City.

Maria Edgeworth.

An example of how landlords of the 1800s regarded their tenants was documented by Maria Edgeworth, an Anglo Irish writer of the period. (Landlady doesn't mean quite the same thing as landlord in this context.)

Although born in London, she spent most of her adult life on her family's estate in Edgewardstown, Co Longford. Her many letters describe another aspect of life in Ireland in the 1830s. She wrote the following about her last novel *Helen* on the 14th February 1834. "It is impossible to draw Ireland as she now is in a book of fiction—realities are too strong, party passions too violent to bear to see, or care to look at their faces in the looking-glass…Then I shall be ready to join in the laugh. Sir Walter Scott once said to me, 'Do explain to the public why Pat, who gets forward so well in other countries, is so miserable in his own.'"

On one hand, Maria was forward thinking in many ways promoting equal rights for women in her local area. She worked hard to improve the living standards of the poor in Edgeworthstown and was involved in efforts to provide relief during the famine. She provided schools for the local children of all denominations. But then Maria Edgeworth showed a remarkable ability to be on both sides of an argument at the same time. "I quite agree with you, as you do with my father, in the general principle that according to the British Constitution the voters at elections should be free, that the landlords should not force their tenants to vote. But a landlord must and should and ever will have influence, and this is one way in which property is represented and the real balance of the constitution preserved. My father in fact always did use the influence of being a good kind landlord, as well as the favour of leaving a hanging half-year in their hands.

I never knew him in any instance revenge a tenant's voting against him, but I have heard him say, and I know it was his principle, that he was not bound to show favour or affection to any tenant who voted what is called against his landlord. The calling for the hanging-gale may, in this point of view, come under his principles, as it is only the withdrawing of a favour - the resumption of a landlord's right."

The hanging gale was a system where a landlord would allow tenants a six-month grace period on payment of their rent until the crops were sold. Maria Edgeworth's justification for landlords influencing their tenants voting rights by withholding the hanging gale was only a small step away from justifying evictions. This type of moral equivocation would provide Francis Spaight with many of his passengers on board *the Thetis*.

The immigrant voyages to Quebec City in the early 1800s were made in the summer months. Winter voyages to Quebec were never undertaken as the St. Lawrence River freezes over every year, making passage impossible. On the return voyages, ships had to leave Quebec City before the winter ice locked in the Gulf of St. Lawrence. *The Thetis* arriving back to the Shannon Estuary in late November 1834 was at the end of the sailing season.

Let's take a step back in time in Limerick to April 1834:

View of the Customs House, Limerick. Painting by J Newman.

This picture reproduced in the previous page is of the former Custom House in Limerick which, in the 1800s, was on the bank of the river. There is now a park in front of building. In the eighteenth century, the Custom House was the centre for the Revenue Commissioners and the Customs Collector. Spaight's ships, including *the Thetis*, had to clear customs here before departing for Canada and report there on return to Limerick. The building is now beautifully restored and houses The Hunt Museum.

In 1834, intending passengers from Limerick to the Americas, bought their tickets as this notice describes.

Apply to the owner Francis SPAIGHT Esq. or to Messrs. MULLOCK & Son, American Passengers' Office, Arthur's-Quay.
– Newspaper notice dating from 1830. *Limerick Chronicle* and others.

Mullock & Sons are still operating as shipbrokers on the Shannon Estuary in Limerick and Foynes.

Kevin Hannan, in *Limerick: Historical Reflections,* described Limerick as looking "like a thriving tourist centre though the people who created that illusion had no money to spend. Large numbers of strangers burdened with their personal belongings were to be seen here and there in the city..." Intending passengers always had to wait in Limerick before departing. It was only on very rare occasions that ships departed on the day advertised.

Having booked a passage on *the Thetis,* passengers' names were entered on a manifest by Spaight's clerk, Sam Evans. Passengers, had to enquire at the pier whether the ship was going to sail that day. When at last, the go-ahead was given, the immigrants rounded up their family members from the lodgings, collected their belongings and made their way to the docks. Their tickets were checked against the manifest to prevent stowaways who were a constant problem for ships' captains. Many were very young boys who had neither family members nor provisions with them. A copy of the details of the passengers was given to Joseph Younghusband before departure as he would need a list on arrival in Quebec City. But very few of these passenger lists on ships from Limerick to Quebec have survived. The exceptions appear to be the lists of those who lost their lives when their ships were wrecked in the middle of the Atlantic. In contrast in the 1820s, ship captains arriving in New York had to swear an affidavit as to the names of the passengers on board. Lists of passengers arriving in New York in the 1830s are readily available. This is one small excerpt of a passenger list from the brig *Hibernia* which had arrived from Dublin in March 1820, and includes the captain's affidavit:

DISTRICT OF NEW YORK – PORT OF NEW YORK

I, J. G. Watheling do solemnly, sincerely and truly Swear that the following List or Manifest of Passengers, subscribed with my name, and now delivered by me to the Collector of the Customs for the District of New York, contains to the best of my knowledge and belief a just and true account of all the Passengers received on board the Br. Brig Hibernia whereof I am master from Dublin. So help me God.
Sworn to, the 13 March 1820. Before me, J. W. Rearney, D. C.

    1 Mrs Hunt 60 female Spinster Ireland New York
    2 Edw. Croswaith 24 female Spinster Ireland
    3 James Croswaith 22 male Clerk Ireland
    4 William Clark 34 male Gentleman Ireland
    5 T. Daley 27 male Gentleman Ireland
J. G. Watheling (Master).

– *National Archives and Records Administration, Film M237, Reel 95, List 150. Transcribed by Judy Caine a member of the Immigrant Ships Transcribers Guild 21 February 2005.*

Passengers had to source accommodation in Limerick in cheap lodging houses for a week or two because of the uncertainty of sailing times. There were hundreds of lodging houses catering for intending immigrants in the old parts of the city. At the same time, the new

Georgian Quarter was being developed in Newtown Perry, stretching from O'Connell's Street to O'Connell's Avenue and covering an area of three square miles. Large three storey houses were being constructed for Limerick's elite and merchant class. Construction work was happening everywhere in Limerick. A new bridge across the Shannon, now known as the Sarsfield Bridge, was nearing completion in 1834. The contrast between the merchant class and people thronging the streets waiting for passage to Canada could hardly have been greater. And there were people even worse off who could never have afforded the fare to cross the Atlantic to the Americas. A Scot recorded the circumstances of the poorest of the poor in Limerick at the time. Henry David Ingles wrote a series of travel books called *Ireland in 1834*, under the pseudonym Derwent Cooper.

"I know of no town in which so distinct a line is drawn between its good and its bad quarters as in Limerick. A person arriving in Limerick by one of the best approaches and driving to an hotel in George Street will probably say 'What a very handsome city this is' while on the other hand a person entering the city by the old town and taking up his quarters, there a thing indeed not likely to happen, would infallibly set down Limerick as the very vilest town he had ever entered."

He described the inside of these homes as being filthy and lacking furniture. They had no beds to sleep on, instead they had to content themselves with bundles of straw and mats. Many of the inhabitants were malnourished, weak and emaciated. There was no escape to the Americas for these unfortunates as the cost of the passage was way beyond their means.

"…the inmates, were some of them old, crooked, and diseased; some younger, but emaciated, and surrounded by starving children; some were sitting on the damp ground, some standing, and many were unable to rise from their little straw heaps. In scarcely one hovel, could I find even a potato…"

A decade before the Great Famine, Inglis recorded that many people were starving to death in Limerick City.

'In one which I entered, [I saw] two bundles of straw lay in two corners; on one, sat a bed-ridden woman; on another, lay two naked children,—literally naked, with a torn rag of some kind thrown over them both….. In a slippery with damp, I found a man sitting on a little sawdust…he was naked: he had not even a shirt.'

And then there was a riot in 1830 in Limerick.

It seems the signal to act was an assault on a convoy of five car-loads of oatmeal which had just crossed Thomond Bridge, heading through the Old Town in the direction of Locke Quay. This occurred between 7 and 8am. The people took most of the stock before constables succeeded in protecting four car loads of it for the proprietor.
– Limerick Food Riots, 25th June 1830.

Henry David Ingles wasn't slow about attributing blame for the appalling conditions so many people in Limerick had to live: "Why should Lord Limerick in Ireland be exempt from the duty which Lord Limerick in England must perform. Why under the same government should men be allowed to starve in one division of the empire and not in another."

Edmund Pery's younger brother, William, was a leading figure in the Church of Ireland. His son, Edmund Henry Pery was made Earl of Limerick in 1803 as a result of his support for the Act of Union. He is probably the 'Lord Limerick' Ingles is referring to. And unlike the Spaights, he was an absentee landlord, residing in London

On that spring day, in April 1834, men, women and children lined up on the pier in Limerick looking at *the Thetis* as it was being towed from its mooring in the river onto the quay side. *The Thetis* is reported as having 217 immigrants on board when it arrived in Quebec City. That doesn't mean that all 217 boarded in Limerick. If there were official checks in Limerick on the numbers, then only 160 people should have boarded the ship. *The Thetis* could have collected more passengers in Kilrush as she passed along the Estuary. Most of these ships made two trips to Quebec every year but often the second journey had no passengers on board. Why didn't they carry 100 passengers on each trip instead of overloading the first voyage so dangerously? One of the reasons was the seasonal nature of the work available in the Americas. And many of the intended passengers understandably preferred the earlier calmer sailings in the summer. But much of the available work wasn't seasonal so perhaps the answer lies elsewhere.

When taking on supplies or passengers, *the Thetis* was pulled into the pier from its anchorage in the river. The ship was 'warped' into the dock from its mooring. The sailors tied a line to the pier and wound it around a capstan, a cylindrical pulley. They pushed the capstan bars around, winding the rope tighter and pulling the ship into the pier. When the ship was moored, the gangways were lowered and the first passengers boarded *the Thetis*. When they were shown where they were to spend the next forty days below deck, was there shock and horror or did they know what to expect? Many of the passengers had family members who had preceded them to America who would have told them about the conditions on board.

Or did they feel like fifteen year old Ralph Rover on *the Arrow* as they left England to sail to the South Pacific in R.M. Ballyatine's *Coral Island*:

"It was a bright, beautiful, warm day when our ship spread her canvas to the breeze and sailed for the regions of the south. Oh, how my heart bounded with delight as I listened to the merry chorus of the sailors while they hauled at the ropes and got in the anchor! The captain shouted; the men ran to obey; the noble ship bent over to the breeze, and the shore gradually faded from my view; while I stood looking on, with a kind of feeling that the whole was a delightful dream."

The Thetis left Limerick just after the turn of the full tide. Ships could depend on the current from the Shannon but the prevailing southwesterlies were a problem as they can blow directly up the Shannon. If the winds were unfavourable, the ship had to be towed by rowing boat into the outgoing tide. When towing wasn't possible, 'ketching' was another option to bring ships into the current. This was a laborious process where the rowing boat brought a small anchor forward of the bow of the ship. The crew pulled the ship forward against the anchor. This operation was repeated until *the Thetis* could avail of the current and wind.

Continuing the account related to George Greaney in 1898, grandson of Dick Greaney, who was an apprentice seaman on *the Thetis* in 1834.

Younghusband was not a bit happy. This was his first trip as captain with passengers. He had served as first mate on previous voyages to Quebec with hundreds of immigrants but now he was in charge and he was nervous. His first mate, Connors, was a good man and very experienced. However, we were shorthanded as few seamen wanted to sign on these overcrowded trips. Three of the skipper's regular crew men had not signed up preferring the shorter runs to Liverpool and London. We were down to the bare minimum of fifteen men, and two of us were mere boys. I know he argued with Spaight about the numbers on board but had been outmanoeuvred by him. Frank Spaight had promised him extra money if everything went to plan. To be fair to Spaight, he had always kept his promises. And then there was the small matter of the tobacco. I wondered had Spaight provided Youghusband with funds to buy four cwt of tobacco in Quebec and deliver to his contacts in North Kerry. Younghusband reckoned he could probably manage to get six cwt of tobacco for the money and sure what Spaight didn't know wouldn't hurt him. Younghusband and Connors had other customers for smuggled tobacco. Between them, they had enough money to buy another two cwt of contraband. Four cwt of tobacco for Spaight and four cwt for themselves would just about compensate for the unpleasant journey ahead. The captain had to tell us in the crew some of this to make sure of our loyalty.

Younghusband reckoned if Spaight kept his promise about the money and the tobacco smuggling, a month of discomfort going to Quebec might be worth putting up with. Still, he wasn't happy. You only had to look at him as we left Limerick. How was he to keep over 200 people under control. He had firearms hidden in his quarters, but I'm sure he dreaded having to use them. Trying to explain why shots were fired to the Quebec authorities was a nightmare he didn't want to think about. The least attention The Thetis caused the better. All he wanted was a quiet life. Joe Younghusabnd wasn't the worst captain I had sailed with. He shouted a bit in that strange accent of his. They said he was from Liverpool, but others could hear an English West Country burr when he issued orders. He left most of the orders to Connors. Now he was someone you did not want to cross – Connors that is I Never trusted or liked him.

As this was some six years before the calamitous collapse of the potato harvest, the passengers on *the Thetis* were not as malnourished as immigrants during the worst of the Famine years of the 1840s. Being stronger and in good health, they were better able to endure the long voyage.

The passengers might also have better informed about than one might assume. There was a book called *Hints of Emigration to Upper Canada (Especially addressed to Middle and Lower Classes in Great Britain and Ireland)* by Martin Doyle, (Third Edition 1834). This was a comprehensive guide for people intending to emigrate to Canada. The first two editions (according to the author) sold out their print run of 5000 copies each. Even 180 years later, it is very readable and must have been a valuable source of information to intending immigrants, particularly, as it is generally upbeat and encouraging. He says in his corrected third edition, "I cannot in my conscience say 'Do not go, to the peaceable and industrious father of a family anxious to leave the waters of national strife and to trust himself to the less terrifying billows of the ocean to be wafted on their bosom from the land of his forefathers.'"

In the book, he lists the store of food an emigrant should take:

4 stone of oatmeal 4 stone of cutlings for gruel 4 stone of sugar, ½ pound of tea, 4 stone of butter, 20 stone of potatoes, and a few dozen eggs, which should be well greased to exclude the air and consequently preserve them fresh.

I must add a quart or two of whiskey for emergencies.

I have thus endeavoured to supply the Emigrant with the most necessary points of information to guide to economical convient and prosperous Settlement I have presented him with a concise and cheap book. Were it dearer he might not wish to buy it; and were it longer he might not wish to read it.

– Hints of Emigration to Upper Canada, Martin Doyle.

It is doubtful whether the immigrants from Limerick to Quebec could afford either the quantity or variety of food listed above. They'd hardly have been emigrating if they had that kind of money. And how would a family of two parents and two children store that amount of food on board *the Thetis?*

Francis Spaight charged the passengers £3 for the voyage. With an inflation multiplier of 25, £3 is the equivalent of £75 in 2021. A family of four paid £300 in today's money. This was a huge sum for 'paupers' as Spaight referred to them. But there was another bonus for Spaight. Because of the 217 passengers and their combined weight, there was no need for as much ballast. Allowing for the children, the weight of the passengers and their meagre belongings must have been around 15 tons. The passengers were accommodated between decks, 'in steerage'. Wooden planking had been placed over cross beams by carpenters brought in to build temporary berths along the sides of the hold. The only means of ventilation was through the hatches, and in stormy seas, the hatches could be kept battened down for days.

In Edwin C. Guilllet's *The Great Migration*, he describes the accommodation on board. "In the regular emigrant ships there was usually a steerage of about 75' long by 20' or 25' wide and 5 ½ high. On either side of a 5' aisle were double rows of berths made or rough planks and each berth designed to accommodate 6 adults was 10' wide and 5' long. Four rows of 13 berths might therefore hold 312 people while the 5' aisle was congested with their baggage, utensils and food."

Before *the James* left Limerick Port in early 1834, an official was appointed to assess the seaworthiness of the ship.

The poor condition of many of the emigrant ships led to calls for improvements in shipping safety. The Corporation of Limerick were so alarmed about the unseaworthiness of some of these emigrant vessels that on April 12th, 1834, they appointed William Vokes as a temporary Inspector of these ships prior to the arrival of Mr Richard Lynch, R.N, the official Emigration Agent appointed by London. Vokes, a policeman, knew nothing of ships. The ships sailed on April 16th. Richard Lynch, a qualified mariner, took up his appointment in May. The Limerick Star newspaper of June 10th reported the shipwreck of the Astrea with the loss of 240 lives. On the 28th June, Lloyds Shipping Intelligence reported the loss of the James. Only eleven of the 241 souls on board survived the shipwreck. Local newspapers reacted to the disasters with fury. The Limerick Herald described the James, a 50 year old hulk, as being a "coffin" to all those who died on her. And so a phrase coined by a Limerick journalist entered the lexicon of the English language.
– *Limerick Archives.*

It was said of *the James* at the time "to have been near half a century old. What a crazy hulk was appropriated for human freight and no wonder she was destined to be the coffin of so many." 230 people drowned on board *the James* as she crossed the Atlantic. Eleven people survived the sinking of *the James* which included the captain William Laidler, his brother Robert and six other crew members. The passengers on *the Thetis* knew how dangerous the voyage could be. They must have known about the losses at sea and the risk that they were taking. The news of the loss of *the James* and *the Astrea* might not have arrived in Limerick before *the Thetis* set sail but there were many other ships that never arrived in the Americas prior to 1834. Henry Downes was a surgeon on the board *the James*. He survived the tragedy so it's worth quoting his account of the sinking of the ship.

To the Editor of the Quebec Gazette

Sir, – Allow me the liberty of intruding on your space with a more accurate detail of the circumstances connected with the loss of the James, which was rather imperfectly given in yesterday's Mercury. We sailed from Limerick on the 8th April, with 251 passengers and a crew of sixteen. On Friday the 11th, we put out to sea, where, after a few days, from heavy gales &c. we experienced nothing but a series of mishaps, having carried away our topmast, studding-sail boom, jib boom, main sail, foresail, and yard. On Sunday the 25th, at six A.M. they set about pumping the ship out, but were not long thus engaged before the pumps were found to be choked by the passengers' potatoes,........ filling the pump wells, and preventing the possibility of working the pumps, ...

Finding the water to increase to an alarming extent, and a gale from the N.W. springing up, with a heavy sea, the ship straining very much, we had recourse to the expedient of baling her out from the fore hatch with buckets and a provision cask made fast to a tackle........... About four o'clock P.M. she shipped a sea, which carried away the lee bulwarks, and was soon struck by a second still heavier, with the force of which she listed, canting her ballast, and never returned to an erect position.......The passengers crowded into the skiff while she was within the long boat, and by this means made it difficult to lower the latter, which, when drawn from the after-chock, came against the stancheons.......

At half past six we lowered the jolly-boat, in which eleven of us were picked up by the Margaret, of Newcastle, Captain Wake, to whose kindness and humanity since we are indebted for our preservation. The persons saved are – Captain Laidler ; Robert S. Laidler, his brother ; Henry Downes, surgeon ; Thomas Enwright, carpenter ; James Cook, seaman ; Peter Lilly Wall and James Clarke, apprentices, with Mary Hastings ; Andrew Young ; James Shehan and Edmund Curry, or Cody, passengers.

Your obedient servant,
Henry Downes,
Surgeon of the James
Quebec Gazette

Some reports say that Lt. Lynch had also inspected *the Astrea* in 1834. But as *the Astrea* left Limerick on the 15th April before his appointment in May 1834 that seems mistaken. The ship was wrecked on the 8th May 1834. Of the 200 passengers that had departed from Limerick on *the Astrea,* only three survived. The wrecking occurred at Loran Head, five miles east of Cape Breton near the entrance to the Gulf of St. Lawrence. One of the survivors, Jerome O'Sullivan, wrote to his sister Catherine of his experience that night.

Charlotte's Town, Prince Edward's Island
May 16, 1834

My Dear Catherine,
I suppose you have seen by the papers the melancholy announcement of the wreck of the barque Astrea, with the loss of all lives on board with the exception of three, the carpenter, a seaman, and your humble servant. I won't detain you by detailing the horrors of a shipwreck; suffice it to say, we struck against a rock at two o'clock, on Thursday morning, May 8th, and were dashed to pieces in less than twenty minutes.

The dangers of the timber ships were well known but it wasn't until 1839 that the issue was addressed by the British Government. A Select Committee was "appointed to inquire into Shipwrecks of Timber Ships and the Loss of Life attendant thereon". The opening page of the Report recorded that in 1834, twenty one timber ships were lost, in 1835 forty nine ships, in 1836 seventy one ships, in 1837 thirty two ships and in 1838 sixty six ships were lost either on the shore or at sea. Given that number of shipwrecks, it's extraordinary that people kept boarding these ramshackle ships in various ports around the Irish coast. This report mentioned the infamous *Francis Spaight* and the sacrificing of the young cabin boy, but it also recorded six similar incidents of cannibalism in 1837 and 1838 on board timber ships. Crews on board the *Earl Kellie, the Caledonia, the Dryden,* the

Earl Moira, the Anna Maria and the Frederick sacrificed a member of their ship's compliment to ensure their own survival. And there were dangers even before *the Thetis* left the estuary.

The Henrietta Sloop, of Ballylongford, was boarded last Friday morning, in the mouth of the Adare river, on her passage home from Limerick, by six men, armed and with their faces painted, who ordered all the passengers up on deck, and rifled the persons of every one of them, carrying off a good booty. While engaged in this daring outrage, the ruffians presented firearms at the heads of their victims, threatening instant death in the event of resistance. They also went below and searched the cargo, consisting of groceries, woollens and mercery, and plundered a bale of silk handkerchiefs, muslins, laces, and shawls.
– *Reading Mercury 12th December 1831*

Gerard Curtin, a Limerick historian, provided a detailed analysis of the passengers who were drowned on board *the James*. The names and occupations of the passengers are listed by him in his article *A Limerick Coffin Ship*. Of the two hundred and forty seven passengers who set sail from Limerick, 147 were from Limerick, 64 from Tipperary, 21 from Clare, four from Cork and three from Kerry. There were 105 men and 84 women on board. The remaining passengers were children under fourteen years of age.

Most of those on board were farmers or from farming families because they could afford to pay the fare. Gerard Curtin suggests that the majority on board *the James* were single people, young men and women. It's not too big a stretch to say the passengers on board *the Thetis* were from a similar cross section of society. While the majority of passengers were Catholic, many of the family groupings on *the James* were from the Palantine community of Rathkeale and Croagh in Co. Limerick. The Palatines, who were originally from Germany, had settled in the Rathkeale area when Sir Thomas Southwell invited them to farm on his estate in 1709. There were families with names such as Delmege, Starke, Ruckle and Hederman on board who were from the Irish Palantine community. They were hoping to join with the large Palantine populations in New York and Pennsylvania

In 1834, what language did the passengers on board *the Thetis* speak? Irish was spoken widely in rural areas, but English was the main language in the towns and cities. The local newspapers in Limerick, Clare, Kerry and Cork were published in English. In 1834, almost everybody in Munster spoke English, but some were bilingual with Irish as their main language. However, perhaps many could read English, but not Irish. The crews on these ships were from many different countries with English being the common language.

The term 'coffin ship' was not yet in use in 1834. That was to come later during the Famine years from 1840 to 1847. But people were emigrating in large numbers. In the pre-famine era of 1815 to 1845, one million Irish moved to North America. The demand for places to travel to America from Limerick meant many unsuitable ships were used and often had disastrous consequences. By the 1840s, the number of unskilled emigrants increased, outnumbering farmers and artisans. There were large numbers of younger sons and daughters of middling tenant farmers. Often, one son was going to inherit the land leaving emigration as the only viable alternative for the remaining children.

I found so great an advantage of getting rid of the pauper population upon my own property that I made every possible exertion to remove them ... I consider the failure of the potato crop to be the greatest possible value in one respect in enabling us to carry out the emigration system.
– Francis Spaight to House of Commons Select Committee, on Emigration in 1847.

Francis Spaight had a chequered history as a ship owner. As we know, the first *Thetis* was lost in November 1834 on a North Kerry beach. The second *Thetis* was also a brig that was built in Sunderland in 1838. This second *Thetis* was wrecked on the Cardigan Bar in Wales in 1850. Eleven men were drowned despite the efforts of the lifeboat crew. Two other ships, called after their owner, were lost at sea. The first *Francis Spaight* was wrecked in 1835. As mentioned already, it was infamous because members of the crew resorted to cannibalism after the ship was wrecked. *The Jessy* went aground on St. Paul's Island at the entrance to the Gulf of St. Lawrence. Even the much larger ship *the Jane Black* got into serious trouble, firstly in 1842 and then again in 1858.

In May 1829, there was a strange article in the *Limerick Chronicle* in which Francis Spaight had to deny reports that *the Thetis* had been lost at sea. The newspaper said that there had been other such reports by "some malicious and designing person or persons". Clearly, Spaight had upset some of his ship owning rivals. If this was malicious, it was particularly cruel for the families of the passengers who departed Limerick on *the Thetis*. They waited for weeks before they found out that *The Thetis* had arrived safely in Quebec.

A report being in circulation that the brig Thetis belonging to this port (which sailed hence on the 14th ult. with passengers for Quebec) has been lost at sea and having reason to believe that such a report has been circulated by some malicious and designing person or persons in order to injure the character of said vessel, the owner Francis Spaight, hereby informs the relatives and friends of the passengers per Thetis that there is no foundation whatever for such report: nor has there been any authentic intelligence of the vessel since she left the river. The wind has been particularly favourable for her since her departure and he confidently hopes she is now in sight of American land.

And whereas similar unfounded reports have been circulated in more instances than this, of his vessels, measures are taken to discover the author or authors of this malicious fabrications in order to take every legal means of punishing such to the fullest extent that the law will allow. As soon as advices come to hand of the arrival of the Thetis notice thereof will be given in the newspapers.

– Limerick Chronicle May 2nd 1829.

The Native also sank but, as we recount later, this was a deliberate act of sabotage. The brig *Bryan Abbs* was abandoned 800 miles west of Newfoundland on route from Limerick to New York in 1854. *The Inverness* was lost in the Shannon Estuary in 1817. It is not clear why this letter is addressed to Spaight as the owner is recorded as being J. Ferguson. 1817 seems too early for Francis Spaight to be a ship owner. Perhaps he had cargo on board but didn't own the ship.

Dear Spaight,
As I am now in possession of most of the particulars of the wreck of the Inverness I shall detail them to you as follows:-
She went on shore on Wednesday night the 19th mistaking Rinevaha for Carrigaholt and would have got off by the next tide.
    WRECK and PLUNDER of the INVERNESS. 1817
– From Capt Miller of the Police to Mr Spaight, Merchant, Limerick, Kilrush, Feb. 2nd.

Later in 1847, Spaight made one of his infamous comments in Westminster. By then, he had been carrying passengers across the Atlantic for seventeen years.

Most certainly because our ships would otherwise go out in ballast and the result is that whatever we get in the way of passengers is so much gain to us.
– Francis Spaight to House of Commons Select Committee, on Emigration in 1847.

What he meant was that sailing ships were unsafe if they were completely empty. They were designed to carry cargo so if the hold was empty, they were unstable and could capsize. Ballast, usually, stone, was brought on board and spread out across the hold to balance the force in the sails. The stone had to be bought in the port of departure and then disposed of before any cargo was brought on board in Quebec. Some countries were short of stone and were happy to take the ballast. Buildings and cobbled roads across the Eastern seaboard of the USA and Canada were built from ballast which was often basalt. However, in some ports, the stone was considered unsuitable. This stone was dumped overboard which caused problems for ships arriving when access to the ports became clogged with ballast.

The Thetis required ballast to sail safely across the Atlantic. How much easier it was to have some of the "ballast" walk on and off the ship. This "ballast" had to be encouraged to spread out across the deck of the ship during the voyage. Discipline on a long arduous voyage across the Atlantic was very important. A sudden rush by large groups of people to one side or the other of the ship could be dangerous to the stability of *the Thetis*.

This wasn't the only time Spaight made such comments. Treating passengers as a preferable alternative to ballast, in balancing the ship, seems outrageous to us now, but Spaight had his motives for being so public with his outbursts. And yet, he often sent empty ships to Quebec, particularly, on the second annual trips across the Atlantic. Critically, if the ballast was not correctly calculated, the ship would become unstable and liable to capsize in heavy seas mid Atlantic. 200 people on board would weigh less than 15 tons, so stone ballast was also used. And if the ballast weight was wrong, then the ship rolled in a sickening fashion. This could happen even in moderate seas making life in the lower decks for the passengers intolerable.

Frank Spaight had identified the relationship between the timber trade, ballast and immigrants. But was he also involved in the illegal tobacco trade? He was the only one who had the funds and contacts on both sides of the Atlantic to purchase and distribute the contraband tobacco. The circumstantial evidence would suggest that he was not only involved but was one of the main members of a tobacco combine.

All the passengers on board *the Thetis* were not allowed on deck at the same time. Apart from the lack of ballast and the consequent instability of the ship, having that many people milling around on the open deck interfered with the handling of the ship. The crew could not operate the ship safely with dozens of children running around the decks. It is more likely that groups of fifty passengers were allowed on deck at a time for an hour or two to cook food on deck and get some fresh air.

The passengers' fireplaces, upon either side of the foredeck furnished endless scenes, sometimes of noisy merriment, at others of quarrels. The fire was contained in a large wooden case lined with bricks and shaped something like an old-fashioned settee – the coals being confined by two or three iron bars in front. From morning till evening, they were surrounded by groups of men, women and children; some making stirabout in all kinds of vessels, and others baking cakes upon extemporary griddles. Which they were provided was of very bad quality – this they had five days and biscuit, which was good, two days in the week.
– *1847 Famine Ship Diary: The Journey of an Irish Coffin Ship*, Robert Whyte.

For at least 10 hours every day, there were people on deck in all weathers trying to prepare food. The other passengers were kept in cramped, unsanitary conditions with little light for often up to twenty hours a day during the six weeks of the voyage.

Continuing the fictionalised account of George Greaney, from 1898, grandson of Dick Greaney, who was an apprentice seaman on *the Thetis* in 1834.

I couldn't stand it. Hundreds of people on board taking up every nook and cranny on board the ship. The smell and noise were terrible. After the seasickness passed, the arguing and fighting began amongst the passengers. And mainly the arguing was over nothing at all.' You took my place, my food, don't push me, leave my daughter alone ... babies cry, they can't help it.' Everybody was so bored during the long days and weeks crossing the ocean. For the crew, it was miserable. We couldn't do our work properly because there was no room to haul sails to change tack. Trying to get an accurate dead reckoning was hard with people everywhere. And worse were the passengers who thought we were their servants. 'Hey, boy, get this or get that...' I hated it. I joined the Thetis to work as a sailor on a cargo ship not as a dogsbody. But there was one place where I could find a bit of peace. When the Captain gave the go aloft call, I was first up the rigging to the spars, but always the last to come down. When the weather was good, I could spend hours aloft and sometimes stayed until midnight. With the help of some of the older crew, from my vantage point, high above the teeming deck, I was soon able to identify the main stars and even the path of the planets. The captain knew what I was up to, but didn't seem to mind too much, telling me to watch for the North Star and tell him its height over the horizon. Ah, it was so peaceful up there compared to the mayhem below.

Going to sea as a member of the crew on board *the Thetis* was a choice many young men saw as an adventure and a way of seeing the world outside Ireland. Crossing to Quebec on *the Thetis* 'in ballast', and returning to Limerick with a cargo of timber was perfectly acceptable, but having to share a small ship like *the Thetis* with 200 frightened immigrants for seven weeks was not.

Many captains on these trans-Atlantic voyages had a strict routine for steerage passengers to follow, including times to sleep, and when meals could be prepared. There was often a list of work to be done, particularly concerning hygiene. Bedding had to be brought on deck regularly and shaken over the side of the ship to get rid of the lice. And this makes sense, as diseases such as cholera and typhoid could decimate a ship.

William Golding's trilogy *Rites of Passage* is about a voyage to Australia in the early 19th century. In Golding's account, Captain Anderson has his standing orders displayed in a glass case for the passengers to read.

"Passengers are in no case to speak to officers who are executing some duty about the ship. In no case are they to address the officer of the watch during his hours of duty unless expressly enjoined to do so by him."

One of Golding's characters, Mr. Colley is summarily dismissed from the quarterdeck by the captain, "passengers come to the quarterdeck by invitation ..... go forward if you please and keep to looard".

By far the greatest suffering and distress and the most persistent and degrading embarrassment for those in steerage centered around the privy, or water closet, that most necessary of all facilities wherever people in great number congregate. – Paddy's Lament, Thomas Gallagher.

The Thetis had very limited toilet facilities. At most, there were two privies or water closets located on the open deck to serve the needs of over 200 passengers. In bad weather, when the passengers were confined to steerage, sometimes for days at a time people had to go where they stood or else use overflowing chamber pots. Women often tried to get a degree of privacy by using the lower 'orlap' or storage deck as a toilet. This presented further problems as this waste was difficult to remove from the lowest part of the ship. Even more gruesomely, this orlap deck was often infested with rats. It is no wonder cholera, typhoid and dysentery spread so easily. The overpowering smell in steerage on board *the Thetis* for the six weeks voyage was unbearable. Very often, chamber pots were offered for sale on the quay sides before ships departed. At every available opportunity, the contents of chamber pots were disposed of over the side of the ship. Years later this most basic of needs was still not met.

"*You have stated that, after getting to sea, the two privies on deck were destroyed?*"
"*Yes. they were only put up temporarily . . . the day before she sailed.. .*"
"*And that there were none below?*"
"*Yes. None below.*"
"*What was the remedy?*"

"There was no remedy . ."
"In consequence of that there was a very bad smell below?"
"You could not stand below." –
– Testimony of Mr. Delany Finch, Minutes of Evidence Taken Before the Select Committee on Emigrant Ships, 1854.

Steerage in the mid-19th century typically consisted of the area immediately below the main deck of a sailing ship. The ceiling height of the between-decks was usually 1.5m. On ships like *the Thetis* many people spent the entire voyage unable to stand upright. The bunks made of rough boards, like pallets, were set up along both sides of the hold.

A cabin passenger describes the situation on a vessel where 180 people were confined in a dark space "not much larger than a drawing room. I popped my head down for a minute or two, but the smell was too powerful for my olfactory nerves - children crying, women screaming; butter, biscuit treacle, herrings, & potatoes, all rolling from side to side, made up a scene of misery & confusion such as I never saw before."
– Packet Ship Gladiator (1846).

In Robert Whyte's diary, he describes a day-by-day account of the voyage from Dublin to Quebec in 1847. The suffering of the passengers is dealt with in minute detail. But, *the Ajax*, a boat of similar size to *the Thetis*, had only 100 passengers on board. The conditions were very bad for the passengers in steerage on *the Ajax* but more than double that number were on *the Thetis*.

Light was admitted through open hatchways and partly through skylights in the deck. There was canvas in the hatchways, but during storms and rough seas these often had to be covered. Sometimes those in steerage were not allowed on deck or may only be allowed for a short time.
– Famine Ship Diary: The Journey of an Irish Coffin Ship, Robert Whyte.

A letter from an Irish immigrant to a friend in Dublin in 1848 is very descriptive of conditions aboard these ships during an Atlantic storm.

Ten o'clock the scene below no light, the hatches nailed down, some praying, some crying, some cursing and singing, the wife jawing the husband for bringing her into such danger, everything topsy turvy – barrels, boxes, cans, berths, children rolling about with swaying vessel, now and again might be heard the groan of a dying creature and continually the deep moaning of the tempest.
– Letter from Thomas Reilly to John M Kelly (April, 1848).

The monument to Francis Spaight, as it looks in the Limerick of today. The inscription reads: Earl St. Germans Lord Lieutenant opened the floating dock on 26 September 1853. Francis Spaight J.P. President of Chamber of Commerce. W.H. Hall Mayor. James Spaight J.P. High Sheriff. John Long Engineer

FOUR
Why Quebec?

The Lower town, overcrowded and dirty, had all the noises, sights and smells of a busy seaport, and sometimes fascinated but more often offended its summer visitors: It is withal a complète puddle, noisy and bustling, filled with people of every shade and shape, tribe and tongue. As this quarter is the resort of sailors, Iumbermen, and newly arrived emigrants, it présents a fearful scène of disorder, filth and in-tempérance;
– *Québec City in the 1830's,* W. H. Parker.

Lower Town, Quebec City in the 1830's. Artist unknown (Quebec Historical Society).

In the 1830s, Quebec City was the capital of the province of Lower Canada and had a population of 30,000. This was increased dramatically by a transient population of immigrants and sailors. Quebec City also had a military establishment of 2000 British soldiers. In the 1830s, in Quebec City, there was an eclectic mix of people, soldiers, sailors, immigrants and even tourists. Because, according to Professor WH Parker of the University of Manitoba "the population was further increased by a considerable number of citizens of the neighbouring republic attracted by their curiosity to view the noble fortress". There was a thriving tourist industry in the midst of the misery of migration from Europe. The 'noble fortress', high above the Lower Town, is the Citadel, a military establishment located on Cap Diament or the 'Rock of Quebec'.

This is a street view in the Old Town of Quebec, or, more correctly, *Vieux Quebec*, where some of the houses date back to the 17th century. According to some sources, many of these streets weren't paved in the

Vieux-Quebec as it looks in 2023. (Credit - Paul O'Dowd.)

1830s making many areas almost impassable in winter. Like Limerick, there was development happening in the port area to improve landing facilities. This allowed Quebec City to develop a major ship building industry, with up to 6000 people employed. There were hundreds of ships being launched every year. This industry brought tradespeople from all over Canada to find work in the ship building yards.

The ships arriving in the 1830s would have moored near the docks to unload cargo and where there was also a convenient landing area for passengers to disembark. The basin of the St. Lawrence River at the port of Quebec City was able to accommodate up to 100 ships at a time. The first steamship, *the Royal William*, to cross the Atlantic on steam alone was launched in Quebec City in 1831. And, like Limerick, there was a huge disparity in prosperity between the Upper Town and the Lower Town.

Napoleon Bonaparte's embargo on British trade which began in 1808 lasted until 1812. Britain was unable to access Baltic timber

A view of Quebec looking towards The Citadel. Richard Short (1759).

during the period of the embargo. They looked west for alternative sources of timber, mainly pine. Lower Canada had vast territories of native forests which were largely untouched. The timber trade between Britain, Ireland and Canada was so well established when the embargo ended that it continued long into the 19th century. The St. Lawrence River connects Lake Ontario to Montreal and onto Quebec City which at the time was the head of navigation for ocean going vessels. Trees were cut exclusively during the winter months because it was easier to cut the trees when the sap wasn't running. Large masts up to 30 metres long were cut for the British Navy from the trees in the St Lawrence Valley. These were the most valuable product of the Canadian forests. But *the Thetis* was only 30 m long and couldn't carry timbers of that length. As the masts were needed in the ship building centres of Sunderland and Bristol, it is unlikely that these longer timbers were carried on the brigs originating in Limerick. Pine was the major species, although smaller quantities of birch, oak, elm, ash, and cedar were harvested.

NEW ARRIVAL OF
TIMBER, DEALS, &c.

FRANCIS SPAIGHT is now landing, ex *Thetis* and *Priam*, from Quebec, 1000 Tons of very superior Red and Yellow Pine, Oak, Ash, &c. 10,000 twelve-feet 3-inch Bright Deals, of prime quality; and several Thousand Staves, Spars, Oars, &c. &c. all of which will be sold on moderate terms.

F. S. has also for sale, at his Timber Yard, *Bedford-Row*, a large supply of the Best Crown Memel, in Timber and Plank; Norway Red and White Deals, PARTICULARLY selected for this market; Oak, Ash, Elm, Birch, Mahogany, Laths, and all other articles in the Building line; and at his Stores, *Honan's Quay*, Plain and Fine CONGOU TEAS, Patent Refined SUGARS, Russia Matts, Slates of all descriptions, &c. &c.

July 23.

This newspaper notice advertises the types of timber brought from Quebec on *the Thetis* and *the Priam* which were to "be sold in Limerick on moderate terms". The notice advertises 1000 tons of timber, very superior red and yellow pine, oak and ash, from two ships. This could mean that the Thetis had carried 500 tons of timber from Quebec which seems an extraordinary load for a ship of 220 tons.

The logs were 'squared', making the timber easier to transport down the rivers. 'Squared' meant cutting the bark from the felled logs, leaving a roughly square shape. The timber was floated down the Ottawa River to Quebec City. It was assembled into large rafts on which were living quarters for the men on their six week journey to Quebec City. The port had the exporting facilities and access to the Atlantic Ocean and to Europe. But there were some in the British Navy that disliked Canadian timber. They felt the longer voyage from Quebec in the damp hold of these ships lowered the quality of the timber, making it more susceptible to dry rot. They claimed ships made of Canadian wood had only half the life span of a ship made

Floating rafts of timber in Quebec. North Wind Archives Wood depot on the St Lawrence River near the city of Quebec, years 1850. Colour engraving of the 19th century.

from Baltic timber. However, in 1834 Canada was the only source of large quantities of suitable timber so the Navy had no choice.

Having full loads on both east and west voyages was ideal for the ship owners. Immigrants landing in Quebec City were able to make their way south to the large urban centres of the US in Boston and New York. The journey to Quebec City was cheaper and less subject to tariffs than the trip to New York. And crucially in New York, as opposed to Quebec City, ships from Europe were inspected on arrival. Fines were regularly issued in New York when overcrowding of ships was discovered.

Only ships registered in Britain were allowed entry into the Saint Lawrence during this pre-free-trade period. There were also no anti-immigration laws in Canada. Many of the immigrants in the early 1800s chose to stay in Canada where they were made welcome. In August 1828, Field Marshall John Colborne was appointed Lieutenant Governor of Upper Canada. He is credited with increasing the population of the province by 50% in the years from 1830 to 1833. Colborne set up a system of immigration which brought thousands of settlers from Great Britain and Ireland to Canada. He expanded the communication and transportation infrastructure by building roads and bridges, making access to the hinterland easier. In the 1830s, the Canadian authorities were largely accommodating of the huge number of immigrants from Europe landing in Quebec. Between 1832 and 1835, 1559 ships brought 53,000 immigrants across the Atlantic from Europe to Canada.

The view depicted in the image on the previous page is of the St. Lawrence River, taken from the Citadel looking back towards Isle D'Orleans and Le Chateau Frontenac. *The Thetis* sailed passed the island into the port of Quebec City.

For most immigrants, the reception services provided by the state were largely positive. The provision of free, safe accommodation, food, information, and guidance might be invaluable to newcomers. But the regulatory and restrictive elements of the same services could prove devastating to those who found themselves on the wrong side of state officials.
– *Receiving Canada's Immigrants*, Lisa Chilton, University of Prince Edward Island.

The answer to the question 'Why Quebec?' is because much needed timber was available in Quebec and because immigrants landing there could make their way to Boston or New York.

The shipslists is a website that keeps a record of ships that arrived at Canadian and American ports in the early part of the 19th century. It names the ships, captains, ports of embarkation, and arrivals and departures to the port of Quebec City. There are also testimonies on the website of the trials people endured during the voyages crossing the Atlantic. The first arrival to Quebec from Limerick was in 1819 and for the next five years only one or two ships made the journey. By 1834, ten brigs were travelling to Quebec from Limerick and some of the ships made the trip twice that year. The various arrivals of *the Thetis* to Quebec City from 1827 to 1834 are recorded on the website.

Two ships called *Thetis* sailed regularly, from Newcastle, Hull and Bristol to Quebec City from 1819 to 1824. It's likely that one of these ships was bought by Francis Spaight, as only one *Thetis* is recorded as sailing to Canada from Great Britain after 1824.

In 1827, Francis Spaight's *Thetis* arrived in Quebec City on 16th September, after thirty three days at sea 'in ballast' with Galt as captain. This seems to have been the first trip from Limerick for *the Thetis*.

Final Voyage of the Thetis

In 1828, the *Thetis* sailed to Quebec City from Limerick. The ship left Limerick on the 17th April and arrived in Quebec City on the 15th May captained by Haughton. The record says there were 400 people on board. This number of passengers is almost certainly wrong. 400 people could not have been accommodated on *the Thetis* while crossing the Atlantic. Perhaps this should be 40.

According to the shipslist, *the Thetis* sailed again from Limerick on the 7th August 1828, arriving in Quebec City on the 24th September with one of the Gorman brothers as captain. This refers to either Daniel or Timothy, both of whom were skippers on Francis Spaight's ships. There were twenty six passengers on board. The ship left Quebec on October 15th 1828.

In 1829, on 15th April, with Dan Gorman as captain, *the Thetis* set sail for Quebec with 130 settlers arriving on 19th May. *The Thetis* sailed again to Quebec at the end of July arriving on September 12th with forty two people on board. This journey is recorded as taking forty nine days. The next voyage of *the Thetis* to Quebec was on 26th August 1830 with one of the Gorman brothers as captain. The ship arrived in Quebec City on 13th October, forty eight days later.

In 1831, *the Thetis* sailed again to Quebec on 4th April, arriving on the 4th May with Outerbridge as captain. She was carrying 216 immigrants. This thirty day voyage was one of the quickest on record.

In 1832, *the Thetis* sailed to Quebec on the 11th April with 232 immigrants on board. She is recorded as arriving on 2nd June with Gorman as captain. Strangely, as both Gormans had a reputation for not overcrowding their ships, 232 passengers was double the legal capacity of *the Thetis*.

The Thetis made a second trip to Quebec on the 25th August 1832, arriving on 6th October 1832. This time Joseph Younghusband was the skipper. This was Younghusband's first recorded transatlantic voyage as captain. There is no record of her cargo except to say she was 'in ballast'. While *the Thetis'* second trip in 1832 to Quebec was Younghusband's first experience as captain, it can't have been his first trip across the Atlantic. He must have served as First Mate with either of the Gormans as captains. It's hard to imagine Francis Spaight entrusting a boat to a captain who had no experience of the Atlantic crossing.

There are no records available on the shipslist website for arrivals in 1833 to Quebec City. But a Montreal newspaper, *The Vindicator and Canadian Advertiser* of July 16th, 1833 notes that *the Thetis* captained by Joseph Younghusband had been cleared for departure from Quebec City on 6th July. *The Vindicator* was a short-lived newspaper founded by an Irishman, Dr. Daniel Tracy from Tipperary.

The Thetis sailed to Quebec from Limerick on the 15th April 1834 carrying 217 immigrants and arrived at Grosse Ile on May 22nd. Younghusband was captain once again. The ship was quarantined until the 5th June 1834. Twenty other ships arrived in Quebec City from Europe on the same day. There is the following note on the shipslist website:

"The undermentioned vessels at present laying at Grosse Isle, are detained under strict quarantine, in consequence of their having sickness or death on board during the voyage, or upon their arrival; and no vessel arriving at the Quarantine Station, under similar circumstances, will be permitted to proceed on her voyage to Quebec…*Mary and Brutus*, from Cork, *the Thetis* and *Priscilla*, from Limerick; *William Fell* from Newry; *Recovery* and *Penelope*, from Youghal; *Hercules*, from Annan; *British Tar*, from Portsmouth; *Friends*, from Dublin."

In May 1834, *the Thetis* anchored off Grosse Ile. The passengers remained on board for ten more days. In some cases, when anchored off Grosse Ile, captains had to provide food for the immigrants on board. When released from quarantine, *the Thetis* arrived in Quebec City on the 5th June 1834 leaving the immigrants on Grosse Ile. Allowing time for the carpenters to remove the wooden berths in the hold and then load the timber cargo, *the Thetis* could not have been ready to depart for Ireland before the 15th June. She was back in Limerick by mid July 1834.

Final Voyage of the Thetis

The port is filling with great rapidity. The telegraph signalised this morning, eighty-one square rigged vessels between the port and the Quarantine Station. When these shall have arrived, which may be expected by tomorrow, the number of vessels in port on 15th May, will be about 160, a greater number than any preceding year, at the same date.
– Neilson's Quebec Gazette.

There is a reference in the shipslist saying that *the Thetis* sailed again from Limerick on 7th August 1834 but no arrival date is given, although it must have been around the beginning of September.

On *the Thetis'* second voyage to Quebec in 1834, the captain's name is not listed but as we know from later events, it was Joseph Younghusband who was in charge. This time, there is no reference to passengers or cargo. There is no record of the ship arriving at the Quarantine Station at Grosse Ile on this trip. Presumably, there were

Quebec seen from the opposite bank of the St. Lawrence (c. 1835) Sketch, Public Archives of Canada

no passengers on board. Having to lay at anchor under quarantine for over a week earlier in the year made the trip less attractive for Spaight. Also, seasonal workers were not needed in Boston and New York in the autumn or winter.

An 1835 sketch from the 'Public Archives of Canada' is of the port area of Quebec with the Citadel visible in the top left. There are ships anchored across the harbour underneath the Citadel on Cap Diamente. In the foreground, are two ships, in full sail leaving the port. The smaller of the two ships is a brig similar in size to *the Thetis*. It can't be *the Thetis* as in 1835 she was stuck on Beale Bar being slowly reduced to a wreck by wind and tide.

But *the Thetis* went somewhere on the East Coast of Canada to collect the cargo of timber and, of course, the contraband tobacco. The average time for the crossing at this time of the year was forty to forty five days. This would give an arrival in Quebec in early September. *The Thetis* must have departed around the 15th October 1834 on her last journey to arrive in Ireland by the 30th November. Forty five days was a reasonable time for the journey across the Atlantic this late in the year. This would suggest that the voyage was uneventful, and that *the Thetis* was not seriously inconvenienced by stormy weather.

The departure of *the Thetis* from Quebec City was 'reported by the Telegraph'. Communication by electric telegraph wasn't invented for another twenty years but there was a system of semaphore known at the time as 'The Telegraph'. This involved messaging from headlands on the coast to the ships approaching land. These optical telegraphs were commonplace in Europe. Towers were placed on hill tops and messages were passed from one hill to another using clapper boards or flags.

There was an earlier optical telegraph line in Canada. It ran between Halifax and Annapolis in Nova Scotia. The optical telegraph also went across the Bay of Fundy to Saint John. Except for the towers around Halifax harbour, the system was abandoned in

August 1800. Native Americans had been using a similar method for thousands of years with smoke signals.

The Telegraphs now in operation as far as Grosse Isle, are now under the superintendence of Mr. Watt, by whom they were originally planned and established.... Since they are in operation to Grosse Isle, that meritorious individual, who is a calculator of time and the tides – an unassuming but exact and skilled observer of heavenly phenomena, such as eclipses, spots of the sun, ...immediately turned his attention to the useful purpose of conveying intelligence on the arrival of the ships at the Quarantine station.
– Extract from *'The Shipslist',* about the 1830s.

What was the destination of *the Thetis* on the second trip to Canada? Perhaps, Younghusband found a quieter port to load the contraband cargo away from prying eyes in Quebec City. But then it was spotted in the Gulf of St Lawrence by the telegraph. But let's go back to the beginning: This notice was posted in the Limerick and Clare papers in April 1829.

First spring ship to Quebec

The beautiful new first class ship the Thetis, of Limerick, 600 tons burthen, D. GORMAN, Master, will Sail for Quebec, on the 4th of April next. The Thetis is well known as a regular trader, and as one of the Fastest sailing Vessels out of the Port, having last Spring landed her passengers at Quebec, in the short space of Twenty-Four Days. Her Cabin and Steerage Berths will be fitted up in the most superior manner, and as numbers are already entered, an early application is recommended.

The American Government having this Season advertised for 10,000 labourers to be employed in finishing the Grand Ohio and Chesapeake Canal, Emigrants going out early in the Spring, will be certain of procuring a speedy and lucrative employment.
– *Limerick Evening Post* and *Clare Sentinel (3 April 1829).*

There is nothing in this notice that's accurate. In 1829, *the Thetis* wasn't new. She was at least twelve years old. The records show two ships named *Thetis* making the transatlantic trip to Quebec from England in 1817. It is highly likely that Francis Spaight bought one of these second-hand ships. She didn't weigh anything like 600 tons. Using the dimensions of the wreck on Beale Strand as a guide, *the Thetis* was a brig weighing around 220 tons. And it didn't leave Limerick until the 15th April, eleven days late.

By 1829, *the Thetis* had already made a number of trips from Limerick to Quebec, all of which took between thirty five to forty eight days. Theoretically, a 24-day trip was possible but it would need a following wind from the east for the entire trip. Sailing to Quebec City from Loop Head meant sailing into the prevailing southwest winds, making a regular 24-day trip improbable. On the return voyage, with the favourable wind and current, twenty four days might have been achieved. Some passengers arrived in Limerick prepared for an immediate departure and a voyage of twenty four days. Instead, because having to wait in Limerick for eleven days and then taking forty days to cross the Atlantic, it meant their overall journey was twice the advertised time. This placed an intolerable strain on their already meagre resources.

To suggest that the "steerage berths will be fitted up in the most superior manner" was stretching the truth to the limit. 217 people were to be crammed into a space equal to four single decker buses with lowered ceilings. The head height was never more than 1.6m in steerage. Many people on board had to stoop while in steerage. The berths were timber pallets on both sides of the hold with a narrow corridor between them. The curved timber walls of the hold were rough and constantly weeping. The most superior manner 'steerage berths' were damp, cold and dark for the duration of the forty day voyage. This was a thoroughly miserable experience for everybody on board. Apart from that there were too many people on board *the Thetis*.

AMERICA.

FOR QUEBEC the beautiful new first class Brig the THETIS, of Limerick, John Heughan, Master, Burthen 600 Tons, now lying at Spaight's-quay, will sail for Quebec on the 5th of April next. The THETIS is not 12 months old; she is a remarkably fast sailer, and as she belongs to Limerick and will be continued in the Canada Trade, Passengers will have many advantages in going out in her: The Master being well acquainted in Quebec, will have it in his power to assist passengers in procuring employment. The Cabin and Steerage Births are fitted up in a most superior style.

An early application is recommended either to the owner FRANCIS SPAIGHT, Esq. or MULLOCK & SON, American Passengers' Office, Arthur's-quay.

N. B.—A Surgeon will go out in the Thetis.

☞ The beautiful new Brig AGNES will sail for Quebec and Montreal, on the 15th of May.

During the 1830s, there were many similar notices posted in local newspapers by Francis Spaight and in this one he promises 'a surgeon will go out in *the Thetis'*. Given that there was no suitable cabin accommodation on *the Thetis*, this almost certainly never happened.

Overcrowding of the immigrant ships had become such a problem, that the Government had made several attempts to limit the numbers on board. The Act of 1828 specified that there should be no more than three passengers for every 4 tons the weight of the ship. *The Thetis* weighed about 220 tons which would have allowed 165 passengers not 217. Even at 165 passengers, *the Thetis* would have been dangerously overloaded. There were almost no sanitary facilities on board so overcrowding inevitably led to the spread of infectious diseases like cholera and typhoid. Two privies for over 200 people were the only sanitary facilities and they were located on the open deck. The conditions on these ships were well known at the time. In 1835, William Smith O'Brien MP for Clare addressed the House of Commons during the debate on the Passenger Acts and said that "...five hundred or 20 per cent of the passengers who embarked at Limerick for North America died at sea". In the 1830s while living in Limerick, Smith O'Brien campaigned on the issues of emigration, education, and port development.

William Smith O'Brien experienced his own long sea journey when he was sentenced to transportation for life to Van Diemen's Land for leading a failed uprising in Tipperary. His life sentence only lasted five years before he was paroled and then pardoned. William Smith O'Brien returned to Europe and eventually Ireland.

William Smith O'Brien MP.

But of course, as *the Thetis* was advertised as being 600 tons in weight, overcrowding on board was not mentioned. Some ships having passed inspection in Limerick with regard to the numbers of passengers on board, collected even more passengers on the way down the Estuary from Ballylongford, Tarbert or Kilrush. These passengers were able to board *the Thetis* as Younghusband waited for the turn of the tide. These extra passengers often paid the captain directly without having to involve Francis Spaight.

But to be fair to Spaight, calculating a ship's weight is more complex than it seems at first glance. For example, the historical record of *the Beagle*, which was of similar size to *the Thetis*, give a figure of '242 bm tons burthen' or burden. BM (Builder's Old Measurement) was the method used up to 1849 for calculating the cargo capacity of a ship. There was a mathematical formula which had evolved since the 17th century which was used to calculate a ships tonnage: $(L - (B \times 3/5)) \times B \times B/2 \div 94$, where L is the length and B is the beam width.

Allowing 100' for the length and 25' for the width, *the Thetis* was 293 'tons burthen'. This is still a long way short of the 600 tons claimed in the newspaper notice. Perhaps, Spaight was talking about the displacement tonnage of *the Thetis*. If they had a sealed dock in the boatyard where the ship was built, they could measure the amount of water displaced by *the Thetis*. Then multiply that quantity by 62 which is the weight of water in lbs per ft^3. This is, of course, the Archimedes Principle. But maybe Francis Spaight meant the 'loaded displacement'. This is the weight of the ship including passengers, fuel, water and ballast. It's doubtful he was doing anything more than promoting *the Thetis* as the way to go to Canada. He wasn't the only shipowner to place these wholly inaccurate notices in the newspapers. And in the same newspaper notice, Spaight also warned his intending passengers about his rivals in the emigration trade.

CAUTION- the numerous losses and disappointments
which are so frequently encountered by Emigrants, should
make them most particular in selecting good and well-known
Vessels, and in avoiding all old and bad Ships, though offered
at a reduced rate. The payment of a few additional shillings
for a passage is immaterial, compared with the safety, comfort,
and speedy arrival of a vessel. Strangers are too often only anxious to obtain
Money, which, in the event of any accident or delay, is always difficult, if
not impossible, to get refunded.
– Francis Spaight.

Members of the crew must have carried firearms to prevent mutiny or unrest on board. The lives of everybody on board depended on the crew being kept well fed and rested when off watch. But the question of discipline on board these ships is seldom remarked upon except for gory details of flogging members of the crew. Floggings were almost, certainly not part of everyday life on these transatlantic crossings. A flogged crew member is unlikely to be available for the return voyage from Quebec. Recruiting crews for these Atlantic crossings, however, became very difficult. They were not going to tolerate that type of extreme discipline.

Besides, the crew were kept busy as Richard Dana remarked:

Tarring, greasing, oiling, varnishing, painting, scrapping and scrubbing ..... in addition to watching at night, steering, reefing, furling, bracing, making and setting sail and pulling, hauling and climbing in every direction, one will hardly ask, 'What can a sailor find to do at sea?'.
– Two Years Before the Mast, Richard Dana.

The Merchant Shipping Act of 1786 required the owners of British and Irish ships with a deck of more than 15 tons burden to register with Customs officers in its home ports. Ships' Registers were, supposedly, sent to the Custom House in Dublin. After 1824 transcripts of the Registers of Ships in Irish Ports were sent to London. The records generated between 1787 and 1823 were destroyed in the Dublin Custom House fire in 1922. The purpose of the 1786 Act was to exercise control over the loading of ships with either passengers or cargo. It had no noticeable effect on ships sailing from Limerick and seems to have been ignored by Francis Spaight and other ship owners.

The following is a report about overcrowding of an Irish ship from a Canadian paper in 1823.

We subjoin from a Montreal paper.... a decision pronounced in the Court of Vice-Admiralty at Quebec, relative to the conduct of the master of the brigantine William, recently arrived from Ireland ; having on board contrary to the laws of emigration, no less than 140 human beings, who suffered the greatest misery during the voyage, arising from the crowded state of the vessel.

The passengers were all labourers of the lowest description, and the want of ventilation, the crowded state of the hold, the absence of all arrangements for comfort, cleanliness, decency, or convenience, occasioned a horrible stench; several were totally unprovided with a place to lay themselves down.

Court of Vice Admirality, Lower Canada Monday, 11th August, 1823

Judge Kerr– This information has been preferred against William Norris, master of the brigantine William, of Dublin, for penalties to the amount of £2500 for having taken from the port of Dublin, and brought to Quebec, fifty passengers more than are permitted by law.

Two other well-known ships that carried immigrants from Ireland to Quebec and New York were *the Jeannie Johnson* and *the Dunbrody*, both of which were much larger than *the Thetis*. *The Jeannie Johnson* was (or is) a three masted barque with a displacement tonnage of 518 tons.

The Jeannie Johnson (replica).

Beginning in 1848, *the Jeannie Johnson* carried 2,500 passengers on sixteen voyages from Tralee to America. Most of these voyages were to Quebec, with only two trips to New York. Although substantially larger than *the Thetis, the Jeannie Johnson* carried much fewer people on most journeys across the Atlantic, apart from one voyage to New York, when *the Jeannie Johnson* had 240 people on board. There was also a doctor on *the Jeannie Johnson* on every voyage. The headroom below deck was a comfortable 2m. And to the great credit of the owner John O'Donovan, a Kerry native, and the crew, nobody died on board *the Jennie Johnson* on any of her voyages across the Atlantic.

In 2000, the new *Jennie Johnson* was rebuilt in Blennerville in Co Kerry, and is now permanently moored in the Liffey, suitably positioned near the Famine Memorial and the Immigration Museum. It made one trip across the Atlantic in 2003, which included an emotional visit to Grosse Ile near Quebec City of which more later. It is most unlikely that a replica of *the Thetis* will be built, however a visit to *the Jeannie Johnson* gives a genuine representation of the hardships faced by hundreds of thousands of Irishmen, women and children as they left Ireland and sailed across the Atlantic to the Americas in the 1800s. The journalist and author Tom McSweeney made an impassioned defence of the replica *Jeannie Johnson* in his book *Seascapes*, which was published in 2005. He saw local people queuing to go on board *the Jeannie Johnson* when it arrived in St. John's, Newfoundland, Canada.

They touched the wood of the ship, they talked to the crew. There were tears of emotion. For them this was the recreation of what they had been told about, of Irish arriving on a sailing ship after crossing the Atlantic.
– Seascapes, Tom McSweeney.

Many of the replica ships around the world don't sail on the high seas any longer. One of the most famous of all, *the Cutty Sark*, is positioned on the quay side out of the water in Greenwich. Surely these ships, *the Jeannie Johnson* and *the Dunbrody*, replicas or not, are a fitting way to remember the great immigration of the 1800s. Making these ships available to future generations helps to ensure that we won't forget what happened. And how do we know that *the Jeannie Johnson* won't be put to sea again sometime in the future?

The Dunbrody sailed from Waterford to Quebec City during the famine years of 1840 to 1847. At 176ft, it was 70ft longer than *the Thetis*. *The Dunbrody* on most voyages to Quebec, carried less passengers than *the Thetis*, averaging 150 passengers per trip.

The Thetis and many other Irish ships arrived into Quebec in the years after 1823 with many more passengers on board than they were legally entitled to carry. Did the Quebec authorities just

The Dunbrody in full sail.

give up counting? Because at the same time similar sized ships were arriving to New York from Great Britain and Ireland with no more than 150 passengers on board. Brigs leaving from Limerick en route to Quebec City had regularly more than seventy extra people than they were legally entitled to. Obviously, even by the standards of the time, *the Thetis* was dangerously overloaded

And still ...

We have frequently heard the character of emigrant ships from Ireland declared to be worse than that of those concerned in the slave trade from Africa; the account given by the passengers of the Thomas Gelston from Londonderry, substantiates the opinion. The passengers by this vessel state the number including children to have been somewhere from 450 to 517. They were nine weeks on the passage and suffered much from want of water and provisions.
– Montreal Gazette August 1, 1834.

The newspaper notice of 1828 placed by Francis Spaight included the hope of well-paid employment in America for the immigrants.

"The American Government having this Season advertised for 10,000 labourers to be employed in finishing the Grand Ohio and Chesapeake Canal, Emigrants going out early in the Spring, will be certain of procuring a speedy and lucrative employment."

The mention of the canal in the newspaper notice was true. But whether the Canadian authorities were advertising for labourers to work on the canal is open to conjecture. The Chesapeake and Ohio Canal, known as the C&O Canal, operated from 1831 until 1924 along the Potomac River between Washington, D.C. and Cumberland, Maryland. The Grand Ohio and Cheaspeake Canal company began importing labourers to Alexandria and Georgetown in August 1829. The workers were promised meat three times a day, vegetables, a "reasonable allowance of whiskey", and $8 to $12 a day for labourers and $20 a day for masons. In 1832, the canal company prohibited liquor in a bid to improve the speed of construction. They had to repeal the ban shortly afterwards. In September of the same year, there was an outbreak of cholera in the camps, which killed a large number of workers. As a result, many others left the camps. There were reports of fighting between the various factions, with the Irish being mentioned frequently. Working on the Grand Ohio and

Chesapeake Canal might not have been the most attractive proposition for the Irish arriving on *the Thetis* in 1834. Indeed, many Irish left Canada immediately and headed for Boston and New York to join up with family members.

Ireland wasn't the only point of departure for people crossing the Atlantic. Hundreds of thousands of immigrants left Europe for the United States and Canada in the early 1800s. Guilllet's *The Great Migration* notes that "in 1831 58,000 and in 1832 66,000 left the British Isles for America and in the latter year shiploads for Quebec left 36 ports in England, 18 in Scotland and 21 in Ireland". Guillet deals extensively with overcrowding on all vessels but notes that Irish ships were usually the worst offenders.

The Irish saw immigration as an escape from the poverty of subsistence farming. Many of the Irish who left in the 1830s were encouraged by their landlords to go. Many others, young men and women were looking for opportunity and adventure. 'Go west young man' is a familiar refrain from Westerns, but that was in a wagon train on land not in a converted timber ship crossing the Atlantic. The only thing that was correct in the 1828 newspaper notice, was that Francis Spaight was the owner of *the Thetis*.

FIVE

The Journey down the Shannon Estuary

As slow our ship her foamy track
Against the wind was cleaving,
Her trembling pennant still look'd back
To the dear Isle 'twas leaving.
So loath we part from all we love,
From all the links that bind us;
So turn our hearts as on we rove,
To those we've left behind us.

As travellers oft look back at eve,
When eastward darkly going,
To gaze upon that light they leave
Still faint behind them glowing –
So, when the close of pleasure's day
To gloom hath near consign'd us,
We turn to catch one fading ray
Of joy that's left behind us.

– 'As Slow Our Ship' – Thomas Moore (1779-1852).

Many early immigrant carriers were, in the words of a contemporary newspaper.

The worst of all the merchant ships of Gt. Britain & Ireland; with few exceptions they are very old, very ill-manned, very ill found; & considering the dangers of an early spring voyage to this port from ice & tempestuous weather, it is astonishing that more serious accidents have not occurred.
– Quebec Gazette, June 2, 1834.

After leaving Limerick, *the Thetis* made her way down the river to Loop Head. She had to leave Limerick at low tide as the tide through the Estuary flows at speeds up to 4 knots. Sailing ships couldn't sail against the tide on the Estuary. The prevailing wind direction is from the southwest making progress down the Estuary very slow. There are stretches of the river where *the Thetis* had to tack repeatedly across the channel to make any progress. The distance to the mouth of the Shannon is about 100 km. Even allowing for a most unlikely top speed of 10 knots, the ship could not have reached the Atlantic before the tide turned. *The Thetis* and similar ships anchored off Kilrush for up to six hours to await the tide before starting the Atlantic crossing. Almost certainly, more passengers boarded the ship during this stopover.

Sailing on the Shannon Estuary in the 1830s was relatively safe due in no small part to the number of lighthouses which kept ships in the safe side of the channel. The first lighthouse was built on Loop Head in 1670. A coal burning brazier on the roof of the lightkeeper's cottage was the first light to welcome sailors home. The cottage-lighthouse with its coal fire was replaced in 1802 by a more conventional lighthouse, built by Thomas Rogers. The tower was about the same height as the present tower with four rooms and a lantern. The 3.6 m diameter lantern contained twelve oil lamps, each with its own concave reflector. The reflected light shone through a convex lens of solid glass. It didn't have an intermittent light until years later when a new masonry tower was built in 1854. The Commissioners of Irish Lights was established under an Act of the Parliament of Ireland 1786 but lighthouses were not included until 1810.

Loop Head Lighthouse.

Kilcredane Lighthouse, about 10 km from Loop Head on the Clare coast started operating in September 1824. but it is no longer a working lighthouse. The tower is still in good condition but the light from the tower has been replaced by two white leading lights on the shore. Further along the Estuary, there is a lighthouse on Scattery Island which didn't come into operation until 1866.

The lighthouse in Tarbert was built in 1834 so it would have been seen under construction by the passengers and crew aboard *the Thetis*.

The 28m limestone tower was built to warn ships of the treacherous rock called the Bowline (Bolands) Rock. The Tarbert lighthouse meant ships could now clear the rock and also use Tarbert as a port of refuge before being piloted through the narrows between Tarbert and Kilimer in Clare.

Robert Howard was engaged to build the tower. Howard had the tower constructed by May 1832, but it took two years to add the lantern and the optic. On the last day of March 1834, a fixed white reflected light was 20m above the high water mark.

The Beaves lighthouse near Askeaton came later. Ships travelling along the Estuary in 1834 had channel lights from Limerick to Loop Head. As we'll see later, this was not the case in the Gulf of St. Lawrence. Lighthouse construction in Ireland in the early 1800s was due in large part to the vision and energy of one man, George Halpin. Born in 1779 in Dublin, Halpin was a civil engineer who was responsible for the redesign of Dublin Port. Because he was so highly regarded as someone who could get things done he was appointed, in 1810, as Inspector of Lighthouses. In his forty four years as the Inspector of Lighthouses, he was responsible for the construction of fifty three lighthouses and the rebuilding of fifteen others. But oddly, for such an important and influential individual, and like Francis Spaight, no portrait of him exists. George Halpin and his son, also called George, were responsible for the building and refurbishment of the lighthouses on the Shannon Estuary at Loop Head, Kilcredaun, Tarbert and The Beaves. Halpin didn't just build the towers, he installed the newest lenses and lights. A remarkable Irishman who has been largely forgotten by history. Although instead of a statue or a monument, George Halpin's lighthouses still stand and keep shipping safe around the coast of Ireland.

Tarbert Lighthouse.

Even with lighthouses, the Estuary could be a dangerous waterway as a newspaper report from January 1833 details:

Disastrous Storm at Limerick
On Friday we were visited by a dreadful storm, the wind setting in at about eleven o'clock from the S.W.S. and continuing to rage for the greater part of the night with relentless fury ... About eleven o'clock on a large sail boat which left Labasheda, County Clare with wheat and a quantity of furniture was driven in near Foyne's Island when five women and a boy besides three boatmen who were on board perished............

The Anne, Thetis, Tiffen, Cleopatria, and Felicity of Kincardine had all sailed on Friday morning but were obliged to put into Scattery in the evening ... The Doreus, Lord Abercrombie, Charlotte, Vigilant, Sarah and Betsy were driven from their moorings during the storm on Saturday night and were not heard of on Saturday morning. The Nautilus was driven on shore inside Bolands Rock. ....

Two of the brigs that were in Scattery ran ashore inside Kerevough Point. The Thetis rode out the gale in Tarbert Roads in company with another vessel. All the rest of the shipping that were driven out have put back into the Roads.

This report was taken from the July 16th 1833 edition of the *Vindicator* newspaper of Montreal, presumably acting upon the usual invitation "American papers please copy". The storm mentioned in the report must have taken place much earlier in 1833, allowing for the weeks it took for the news to cross the Atlantic.

Brian Goggin's in his *Waterways and Means* remembers Thomas Ennis Steele, known as 'Honest Tom'. Steele was from the Derrymore Estate in Co. Clare. His work on the Estuary was titled *Practical Suggestions on the General Improvement of Navigation of the Shannon between Limerick and the Atlantic*. Steele, a somewhat eccentric member of the landed gentry, had correctly identified that the hazards on the upper reaches of the river were insufficiently marked. Steele quoted a Limerick pilot, "Often, they make us tremble in our skin, going by there at night, Sir. The channel is so narrow and such a rapid tide, Sir."

However, despite Steele's detailed survey and recommendations very little was done, mainly because in the 1830s there was no single authority in charge of marine safety on the Shannon Estuary.

When *the Thetis* travelled through the Estuary, the passengers had the first opportunity to see the Shannon dolphins cavorting in the bow wave. This 120 strong pod of bottlenosed dolphins has been resident in the Estuary for centuries. They have been accompanying ships and boats out to sea for generations. On the north side of the Estuary is Scattery Island, the site of an early Christian settlement founded by St Senan, which was well known to the passengers of *the Thetis* as a place of pilgrimage.

The Thetis passed through rapids between Tarbert and Kilimer where the Shannon narrows causing turbulence as the tide ebbs and flows. A graveyard in Kilimer has one of the most famous graves in Irish literature. Ellen Scanlan was murdered at the request of her older husband on July 14, 1819. Ellen Scanlan, nee Hanley, was only fifteen years old when she was murdered. She has since become known as the 'Colleen Bawn', an anglicised spelling of *Cailín Bán*.

In the later part of the 18th century, this part of the Shannon was the scene of a different form of transport. Richard Crosbie was the first Irishmen to make a manned flight in Dublin in 1785. In May 1786, he made another flight down the Shannon from Limerick to Tarbert. And 48 years later in 1834, there was at least one person on board *the Thetis* who was old enough to have seen Crosbie's balloon as it passed over Tarbert.

On May 1, 1786, in a letter to the *Limerick Chronicle*, he gave a detailed description of his flight a few days earlier.

As the balloon lifted over the "River Shannon with all its little islands formed a pleasing variety I had before been unacquainted with, I determined to take a drawing of it" ...

Regaining his bearings, now he could see that he was heading west, seeing the Atlantic Ocean in front of him. The winds then pushed him towards Tarbert and he saw the Kerry landscape as "plain chequered like a carpet". It was at this time that the balloon became almost stationary "over a small green field for upwards of half an hour", during which time he ate his dinner and drank a "bottle of wine to the health of my numerous friends on earth".
– Limerick Chronicle.

The Thetis' first stop, four hours after leaving Limerick, was Kilrush in Co.Clare to wait the turn of the tide. In April 1834, the police boarded *the Thetis* while anchored off Kilrush to arrest one of the passengers. John Collins was arrested for raping Johanna Spillane the year before. Collins was taken off the ship and brought to Killarney to face trial. The following day, *the Thetis* resumed the journey to the Atlantic. The passengers on board could see Beale Strand in the distance on the Kerry coast just where the Dooneen Cliffs end. Nearly 100 hundreds year later in 1930, Jack McAuliffe from the village of Lixnaw in North Kerry was a regular visitor to Beale. He wrote the famous ballad 'The Cliffs of Dooneen' while staying in Beale.

You may travel far far from your own native home
Far away o'er the mountains far away o'er the foam
But of all the fine places that I've ever seen
There's none to compare with The Cliffs of Dooneen.

Fare thee well to Dooneen fare thee well for a while
And to all the fine people I'm leaving behind
To the streams and the meadows where late I have been
And the high rocky slopes of The Cliffs of Dooneen.

We'll never know whether Jack McAuliffe knew that the shipwreck on the beach had transported over 1000 Irish men and women across the Atlantic to North America in the early nineteenth century. He was friendly with a local man, Liam Buckley who lived beside the beach. Buckley must have been familiar with *the Thetis* and how it came to be stranded in Beale. The song is so evocative it could have been written by one of the passengers on *the Thetis* as it was leaving the Shannon Estuary.

Continuing the account related to George Greaney in 1898, grandson of Dick Greaney who was an apprentice seaman on *the Thetis* in 1834.

We were anchored off Kilrush the night before we left for Canada waiting the turn of the tide. Captain Joe's humour improved the moment the last pilot left the Thetis. He didn't like many of the pilots because they took control of his ship. He particularly didn't like Jake Fitzpatrick who seemed to take great delight in annoying our captain. Joe called them inland sailors and couldn't wait to get them off his ship.

I remember afterwards while we were at anchor, he stood on the deck and assembled the passengers. He made a speech telling them what to expect for the next six weeks. If they did what he and the crew told them to do then they would all get on fine. They must obey Connors who would be in charge of the first watch and Johnny Dee who would take the second watch. To be honest, he went on a bit too much about cooking and distribution of water. He talked about discipline and keeping the children under control. Some of the passengers listened intently but others were paying no attention. Looking back, I'm not sure Joe Younghusband was listening to himself either. All he wanted to do was to set sail for Canada.

Like all large stretches of water, the Estuary can be dangerous to navigate unless you are familiar with it. The people who knew the Estuary best were local boatmen. For centuries, these boatmen have been making a living as pilots guiding ships through the Estuary. There were two groups of pilots operating in the Estuary in the 1800s.

Visitors from across the Atlantic, brent geese, flying over the remains of The Thetis.

The Western Pilots guided ships on their inbound voyage from Loop Head to Scattery Island. The Eastern Pilots from Limerick guided the vessel upriver from there to the port. The outward-bound voyages were the reverse of this, although the Eastern Pilots could guide a craft for the entire journey but they had to pay compensation to the Western Pilots Fund for doing so. Scattery Island's connection with the pilots began with the settling of families from the Kilbaha area of West Clare.

On the morning of March 13th, 1843, the Kilbaha Pilots observed a de-masted ship, *the Windsor Castle* nine miles off Loop Head. When they boarded her, they saw that nobody had remained onboard. She was on route home to Liverpool from Bombay, now Mumbai. The pilots brought her to Kilbaha. A lengthy court case ensued regarding the salvage rights to the vessel. Eventually, a settlement was agreed which provided each of the families with between £110 and £160 pounds. They used the monies to purchase lands on Scattery Island from the landowner Marcus Keane. They relocated to the island and their descendants continued to provide a pilot service on the western part of the Estuary for over a century.

On Monday three men, two of them pilots, went out from Carrigaholt to bring a vessel into Kilrush. The pilots were so stupidly drunk that they upset the boat and were drowned.
– Clare Journal, Thursday 11th September, 1834.

After leaving the Shannon Estuary, Younghusband didn't turn *the Thetis* immediately southwest towards Quebec. As he looked towards the horizon, he knew the route he was going to take. The journey from Ireland to Canada had been undertaken by countless captains for centuries before 1834. Younghusband had his passage clearly mapped out before leaving the docks in Limerick. The compass bearing was already decided. Almost certainly, he carried a rutter or pilot book giving details of previous voyages. These books were used around the world as aids to navigation. Known as a *periplus* by the Greeks, they contained notes and sketches of coasts which skippers could use to identify their location.

Brig Ornen drawing with pencil (Danish Maritime Museum). This brig which was part of the Danish Naval Fleet from 1842 to 1866 was of a similar size to the Thetis.

As captain, Younghusband carried various papers to ensure safe passage across the Atlantic. He had charts of the Shannon, the Gulf of St. Lawrence and the St. Lawrence River. He had records of his earlier voyages and logbooks. Every day's progress crossing the Atlantic was recorded with the dead reckoning marked on a chart. He also needed bills of sales for Spaight's Canadian shipping agents so he could secure the timber for the return voyage. And probably, he carried cash in some form to pay for the contraband tobacco.

The Great Circle Route is shorter than a direct line between Loop Head and Quebec. The direct route, known as the rhumbline, is a straight line between two points on the globe, whereas the Great Circle Route to Quebec is shorter by several days. However, lack of precise navigation instruments required a different strategy to ensure safe passage.

The Thetis had to travel over 4200 km across the Atlantic from Loop Head to the Gulf of St. Lawrence. As captain, Joseph Younghusband needed to plot a passage across the Atlantic to Canada. The 1834 voyages on *the Thetis* were not Younghusband's first trips to Quebec. By then, the route was well known to him. There were so many ships making the journey from Limerick to Quebec, it was like a shuttle service. It's possible two or more ships travelled in convoy in case of trouble at sea.

The brigs crossing the Atlantic from Europe were capable of top speeds of 10 knots, but because of current and wind the average speed was reduced over long distances. If *the Thetis* tried to take a direct line to Quebec, she would have to sail into the prevailing southwest winds. *The Thetis* had to sail 60° off the wind compared to a modern yacht which can sail much closer, at 40° to the wind. The distance from Loop Head to the coast of Canada is 4200 km and the average time to cross the Atlantic was thirty days, giving an average speed of 3.6 knots or about 150 km a day. Another ten days were needed to sail through the Gulf of St. Lawrence and the St.Lawrence River..

However according to a UCD report by Morgan Kelly and Cormac Ó Gráda in 2014 this was changing.

Our results are striking. For ships sailing ahead of a moderate breeze (Beaufort Force 4, the normal summer wind conditions in the North Atlantic) daily speed increased by around one third between 1750 and 1830 from an average of 4.5 to 6 knots. This increase is not steady but occurs in two bursts: the first during the 1780s when sailing speed improves by half a knot to 5 knots, and the second after 1815. In stronger winds the increase is lower; while in light breezes increase is greater, with sailing speed almost doubling from 2.5 to nearly 5 knots. Ships were sailing faster in light breezes in 1830 than they had been in moderate winds in 1750.

The improvement in the post-Napoleonic period does not appear to be associated with any individual major innovation and is likely to be due to incremental innovations in hull profiles, the design of sails and rigging, and the setting of sails.
– Speed Under Sail, 1750-1830, Ucd Centre for Economic Research Working Paper Series 2017, Kelly and O'Grada.

The Gulf Stream originates in the Gulf of Mexico, and follows the coast of the United States before turning east to cross the Atlantic Ocean. The current is 100 kilometres wide with a maximum speed of 5km/hr. *The Thetis* had to negotiate the effects of the Gulf Stream and the prevailing southwesterly winds. If Younghusband took the direct line to Quebec, because of the prevailing winds, *the Thetis* had to tack across the Atlantic. Tacking is zig zaging back and forth as sailing ships can't sail directly into the wind. She also had to sail directly against the flow of the Gulf Stream making this route improbable. The simplest route was to sail along a line of latitude, perhaps the same as Loop Head, 52° north until *the Thetis* was beyond the influence of the Atlantic winds and the Gulf Stream current. This route brought the prevailing south westerlies on to the beam of *the Thetis* reducing the need to tack so often. And importantly, this line of latitude route made precise estimates of longitude less critical.

The first thing Younghusband did when leaving the Estuary was to set the course. If he sailed a parallel, he would have sailed due west, 270° on his compass. By sailing the 52° parallel he was setting a course that would miss his destination to the north of the Gulf of St. Lawrence. Younghusband knew that when he had arrived at the Canadian coast, he had to turn and sail south. If he set a course directly for the Gulf, he wouldn't know whether to turn north or south when *the Thetis* arrived off the Canadian coast. Aiming higher or lower than the destination was common practice among mariners of the period.

These days, we have marine charts and GPS chart plotters to accurately determine our position at sea. In 1834, Younghusband on *the Thetis* had to rely on very different methods of navigation. And yet the underlying science remains the same. Imaginary lines, latitude and longitude, running east west and north south are used to pinpoint every ship's position at sea. There was another straightforward method of determining a ship's position as documented by Edwin C. Guilllet in *The Great Migration*. "As for the crews they were commonly composed of rum soaked illiterate bear like officers who could not work out the ordinary meridian observatory with any degree of accuracy and either trusted to dead reckoning or a blackboard held up by a passing ship for their longitude……..."

But what's wrong with asking a passing ship what they estimate their position to be? In an era, when dead reckoning was the most accurate navigation available, wouldn't it make sense to check the longitude with another captain? If it was a passing ship with local knowledge, it was the sensible course of action.

Saturday 10th July 1847
We spoke to a wherry which was conveying cattle from Nova Scotia to Newfoundland and learned from the steersman the bearings of St. Paul's Island.
–Famine Ship Diary, Robert Whyte.

And almost certainly, drinking rum was safer than drinking water at the end of the voyage on these overcrowded timber ships.

Latitude lines are parallel to the equator whereas lines of longitude connect the poles. To find latitude, sailors measured the Sun's meridian altitude (mer alt), the maximum height above the horizon at local noon. Tables in the almanacs were used to determine the sun's true altitude from which latitude was calculated.

Finding longitude was much more complicated. Longitude was a major navigation issue, whereas finding latitude was easily established. Finding longitude at sea was a problem occupying navies, scientists, astronomers, kings and governments for centuries. The story of how John Harrison's chronometers influenced navigation and the pinpointing of longitude is documented by Dava Sobell in her wonderful book *Longitude*. Richard Dunne and Rebekah Higgit's *Finding Longitude* is also a valuable source of information. Dunne and Higgit take a somewhat broader view, attributing credit for resolving the search for longitude to many others, including French and Italian watchmakers. However, in 1834, chronometers were still not in general use in merchant shipping and almost certainly not in ships based in Limerick.

David Barrie's book *Sextant* is a day-by-day account of how he crossed the Atlantic in 1973 using the same methods that Younghusband on *the Thetis* relied on in 1834. Barrie's account of his voyage across the Atlantic in an 11m yacht called *Saecwen* is interspersed with the story of the sextant through the centuries. The *Saecwen* was owned by Colin McMullen, a retired Royal Navy captain who was an expert on celestial navigation. McMullen learned how to sail in Waterville, Co. Kerry as a boy. Barrie was keen to learn all he could about navigating from McMullen on the voyage across the Atlantic. Barrie joined McMullen and Alexa Du Vivier, in Portland, Maine from where they sailed to Halifax, Nova Scotia before crossing the Atlantic. "As we sailed into the Atlantic, leaving the land further and further astern, I watched the night sky as I had never done before. I recognized some of the main constellations, and, with the help of a star chart, picked out a few of the fifty-odd 'navigational stars' whose coordinates are listed in the Nautical Almanac. Aldebaran, Alkaid, Alioth, Antares, Arcturus, Capella, Mirfak – as well as the three stars of the so-called Summer Triangle – were all in sight." The three stars making up the imaginary summer triangle are Altair, Deneb, and Vega.

In 1973, The Global Positioning System was in its infancy. It wouldn't become properly operational until 1993 so the crew of the *Saecwen* relied on taking fixes on the sun and stars to determine their position at sea as they stayed along the 42° parallel. This was the same method, in the same sea but some 10° to the south,

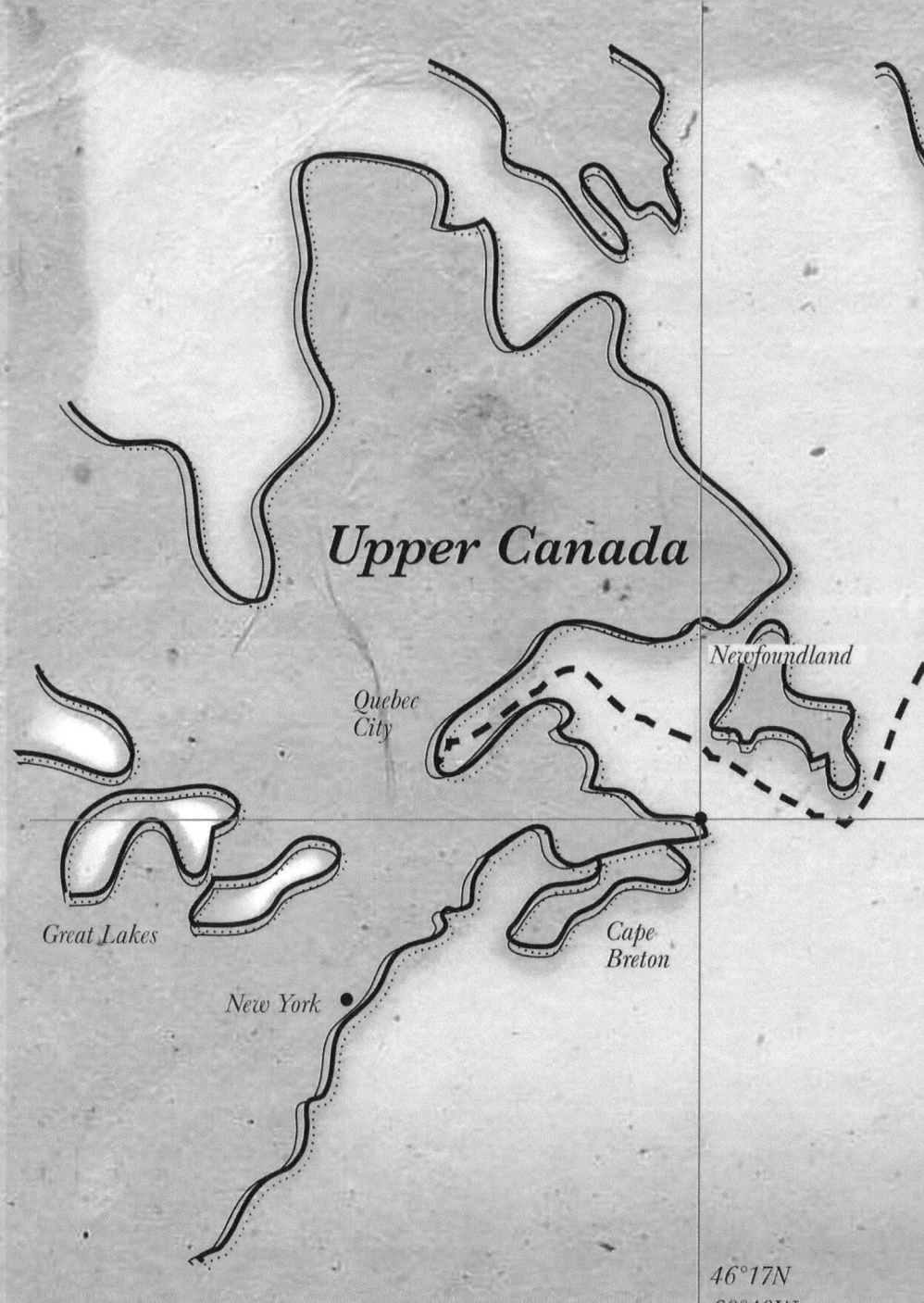

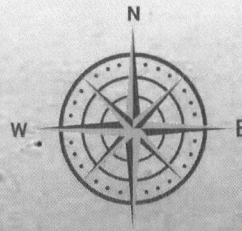

used by Younghusband on board *the Thetis* 139 years earlier. Perhaps Younghusband went west along the 52° line and on the return, journey sailed east on the 42° parallel. "Apart from the sound of the wind and waves all was quiet. Saecwen heeled to the south easterly breeze and began to dip her bows into the Atlantic swell."

But it wasn't all plain sailing on board *The Saecwan*...

Later that morning the wind increased to gale force and we switched down to the smaller storm jib. Even so our speed through the water increased and we were surfing down steep, breaking waves that grew bigger all the time. The speed of change was startling; in the space of an hour or two our world had been completely transformed.
– *Sextant,* David Barrie.

On *the Thetis*, Younghusband would need to know longitude only occasionally on the way to Canada but even then, an approximate fix would do. Ship captains had to know how to use a sextant so they could pinpoint noon at their location. As mentioned already, chronometers in 1830 were prohibitively expensive. Brigs like *the Thetis* sailing from Limerick didn't have them on board. If they did have chronometers on board, where are they now? Surely, some would have survived and be on display in museums?

John Harrison's H4 in 1759 had cost over £20,000 to develop in today's value. Kendall's K1 cost £500 to buy when first available and his cheaper model, K3, cost £100 to produce. By 1834, chronometers could be bought for £60. Chronometers were becoming more affordable but like all new technology, it was taking time for them to be accepted by ship owners. Was Francis Spaight going to spend the modern-day equivalent of £2,000 on a chronometer for *the Thetis*? No, is the simple answer to that question.

There was concern at the highest level about conditions on board ship. In 1834, the King of Great Britain and Ireland was William IV. William had served in the Royal Navy in his youth, in North America and the Caribbean. Before he became King, he was Lord Admiral where he insisted on regular reports on the condition of every navy ship. As an example, he abolished the cat o' nine tails for most offences other than mutiny immediately improving the lives of the sailors. Because of his interest in the Navy, he was known as the "Sailor King". As an aside, George Washington approved a plot to kidnap him, writing:

William IV, in Greenwich.

The spirit of enterprise so conspicuous in your plan for surprising in their quarters and bringing off the Prince William Henry and Admiral Digby merits applause; and you have my authority to make the attempt in any manner, and at such a time, as your judgment may direct. I am fully persuaded, that it is unnecessary to caution you against offering insult or indignity to the persons of the Prince or Admiral.
– From George Washington to Matthias Ogden, 28 March 1782 (archives.gov).

If I'm correct, Younghusband took the simplest route across the Atlantic by 'sailing the parallel'. The navigator on board *the Thetis* would know exactly how many miles each 15° represented. Along the 52° line of latitude, 15° of longitude is about 630 miles. The same 15°

John Harrison's H4 (1759) in The Royal Observatory in Greenwich.

is nearly 1000 miles at the Equator and zero at the poles Christopher Columbus sailed the parallel when 'discovering' America in 1492. Not a particularly strong argument for the practice as Columbus thought he was heading for India.

Staying on a line of latitude makes dead reckoning easier to calculate. Ordinary pocket watches were available in 1830 but their accuracy was affected in extreme temperatures at the equator and in the polar regions. Darwin's *Beagle*, for example, crossed the equator and rounded Cape Horn on their way to Ecuador. This subjected the many chronometers on board to a temperature range of 30° Celsius. The *Beagle* carried 22 chronometers on its voyage to the Galapagos Islands off the coast of Ecuador in the 1830s. By the time the ship returned to London two years later, over half the chronometers were no longer working.

Exploring in unchartered waters made the pinpointing of longitude extremely important. *The Thetis*, on the other hand, was sailing in a temperate zone along well-known trade routes. An ordinary pocket watch, set to local time in Limerick, allied to regular sextant readings should have been sufficiently accurate on the outward journey.

In 1834, Richard Dana in *Two Years Before the Mast*, had an interesting take on chronometers:

> Letter from Captain WILLIAM S. SEBOR, of the Ontario Packet Ship, to Messrs. PARKINSON and FRODSHAM.
>
> "London, Feb. 24, 1832.
>
> "GENTLEMEN—It affords me much pleasure to have it in my power to say, that during the last ten years that I have constantly been engaged in the navigation between this Port and the United States, I have, on board of the vessels under my command, invariably availed myself of sundry Chronometers made by you. They have at all times performed with equal regularity and correctness: and as an additional instance of their durability and perfection, allow me to mention the circumstance, that when on a late voyage to this country one of these instruments accidentally fell from a high shelf, its action did not become thereby in the least disturbed, but, on the contrary, proceeded without the slightest alteration of rate. It is in justice to your superior skill, and already amply-established reputation only, that I have the gratification of stating these facts, the result of personal observation. Wishing you every farther success your meritorious exertions so well deserve, I remain, &c. WILLIAM S. SEBOR."

Letter from William S. Sebor to Parkinson & Frodsham

This was done to determine our longitude: for by the captain's chronometer we were in 25 deg W. but by his observations we were much farther and he had been for some time in doubt whether it was his chronometer or sextant which was out of order. This landfall confirmed the matter and the former instrument was condemned and becoming still worse was never afterwards used.

Lunar distance or lunars is a complex but important means of determining longitude at sea. It remained useful throughout most of the 19th century as many ship owners couldn't afford expensive chronometers. Using lunars, astronomical tables and a sextant, longitude could be found. The fifth Astronomer Royal, Nevil Maskelyne who has been accused by some of obstructing John Harrison's claim to the Longitude prize, was instrumental in publishing the first edition of the *Nautical Almanac*. The almanac contained the positions of the sun, moon and stars on any given day of the year. The *Nautical Almanac and Astronomical Ephemeris* with tables prepared by the German astronomer Tobias Meyer was first published in 1766 and is still published annually. A copy of the *Nautical Almanac* is on every ship of the Royal Navy to this day.

There were others who were very much in favour of chronometers. The letter reproduced on the previous page dates from 1832 and was written was written by a transatlantic captain, William S. Sebor to Parkinson & Frodsham praising the accuracy and durability of their chronometer. Although it should be pointed out, it was published in a short advertising article by the manufacturers.

Nevil Maskelyne was the British Astronomer Royal from 1765 to 1811.

Marine Chronometer by Parkinson & Frodsham, No. 2349.

Nathaniel Bowditch.

Nevil Maskelyne was the major figure of the era in the search for a longitude method at sea. Without his dedication and particularly his insistence on scientific accuracy, the search for a method to accurately determine longitude would have been delayed for years. The lives of thousands of sailors aboard navy and merchant shipping depended on Maskelyne making the right decisions to find a successful and sustainable method of determining longitude.

An American, Nathaniel Bowditch was also important in navigational history as he introduced easier methods for solving lunar distances by using sight reductions. So even when chronometers were available, they were used in conjunction with lunar almanacs for many years. The *Bowditch Almanac* or its successor is still published and is used on all US Navy ships.

James Cooke claimed that "This method of finding the longitude is not so difficult but that any man with proper application and a little practice may soon learn to make these observations as well as the astronomers themselves." That might have been correct for well-educated masters on board Royal Navy ships, but it was not the case for captains sailing from Limerick to Quebec in 1830. They had to rely on a trailing log, a compass, a sextant and almanacs. Established methods of lunar observations and published almanacs were sufficient to fix longitude to the accuracy required to cross the Atlantic but these calculations were difficult and laborious.

More importantly, Younghusband, on board *the Thetis* would need to know where he was along his chosen line of latitude. The tried and trusted method to do this was called dead reckoning. Although, this method could be inaccurate if not carefully carried out as it was subject to so many variables of wind and current. The first mate checked the boat's speed every four hours by throwing a taffrail log on a line overboard. When the log was tossed overboard, it remained more or less stationary while an attached line (marked off with equally spaced knots) was let out behind the vessel for a measured interval of time using a sandglass. The log was then hauled back aboard the ship. The speed of the ship was determined by dividing the length of the line by the time interval. The number of knots hauled in was used to determine the speed of the ship. A knot, a sea mile per hour (1.15 miles) is still used as a measure of speed at sea.

James Cook (Master Surveyor).

For example, on *The Constitution (Old Ironsides)* they checked their dead reckoning every hour using the same procedure. When the log hit the water, the Quartermaster turned the glass. At his "Mark, the seaman would stop the run-out then as he reeled the line back in, the Midshipman noted how many knots had run out."

The crew on *the Thetis* used dead reckoning to measure the distance travelled every day. By knowing the speed and using the compass bearing, Younghusband plotted his position along each tack, port or starboard to compute the dead reckoning. This method was used on all sailing ships of the age. This mattered when ships were exploring in uncharted waters but for Younghusband on *the*

Thetis sailing a parallel on a well-travelled route, it was not as critical. When the weather allowed, a quick fix (a nautical expression) on the sun, moon or stars to establish latitude, was as much information as required for most of the voyage. Longitude would become important the closer *the Thetis* came to the Canadian coast.

If *the Thetis* stayed on the 52° line of latitude, she could hardly miss Canada. If because of the inaccuracy of dead reckoning, *the Thetis* drifted south to a latitude between 52° and 46° Younghusband wouldn't have been concerned as it didn't matter so long as he stayed north of the entrance to the Gulf of St. Lawrence. Once the effect of the wind and the Gulf Stream had lessened close to the Americas, *the Thetis* could turn south and sail down the coast until the 46° line of latitude was reached at the mouth of the Gulf of St Lawrence. A theory, perhaps, but look at the route taken by *the Margaret Bogle* in the 1820s. The three masted barque sailed an approximate parallel to the southern tip of Greenland.

The Margaret Bogle made several voyages to New York, Quebec and Montreal. One such voyage from Leith (Scotland) in 1826 included a visit to Waterford, Ireland before proceeding to Quebec....

Of the three possible routes from Scotland to New York, the Margaret Bogle took the northern route... Picking up the west Greenland current around the southernmost tip of Greenland, she sailed for Baffin Bay to pick up the Labrador current coming south out of Davis Strait.
– *Genealogy of Ormiston,* Magaret Bogle.

Not 'missing Canada' was easy enough, but Younghusband needed to know when to turn south to the 46° parallel off the Gulf of St. Lawrence to make sure he didn't 'hit' Canada. As well, Younghusband had his pilot's book or rutter to refer to. These pilot books included sketches of the Canadian and Irish coastlines, that had been drawn on earlier voyages by previous captains. Reaching a way point along the 52° line of latitude would tell Younghusband when to turn south. Younghusband might have taken a sighting with his sextant on the sun (mer alt, to use the slang) or the moon and then consulted his almanacs to establish his longitude off the coast of Canada.

Shakespeare knew of another important navigation tool, the North Star, because as he said 'it is an ever-fixed mark'. Also known as Polaris, it is located close to the celestial north pole. Polaris is mentioned in Nathaniel Bowditch's 1802 book, *American Practical Navigator* where it is listed as one of the navigational stars. To locate Polaris, find the two stars at the outer edge of The Plough, Dubhe and Merak. Draw a line from Merak through Dubhe and follow the line straight to Polaris.

O, no! it is an ever-fixed mark,
That looks on tempests and is never shaken;
It is the star to every wandering bark,
Whose worth's unknown, although his height be taken.
– 'Sonnet 116', William Shakespeare.

Younghusband on board *the Thetis* had to be acquainted with the night sky. By knowing the location of Polaris, he could determine the direction of north and by extension south, west and east. If the compass wasn't working Polaris could always be relied on. But the North Star had another useful function. Latitude can also be found by measuring the altitude of Polaris above the horizon. If a navigator measures the angle to Polaris from the horizon as 50° then the latitude is 50° north. This reading is corrected using an almanac to give the latitude of the ship accurate to less than one kilometre. If a sextant wasn't available then by counting the number of fists with an arm outstretched from the horizon to Polaris, a rough estimate of latitude could be obtained.

Then the trav'ller in the dark,
Thanks you for your tiny spark,
He could not see which way to go,
If you did not twinkle so.
– *Twinkle Twinkle Little Star* by Jane Taylor (late 1700s) and set to music by Mozart.

Although appearing to the naked eye as a single point of light, Polaris is a triple star system, composed of a *yellow supergiant*, Polaris Aa, in orbit around a smaller companion. These two stars are in a wider orbit with a third smaller star. Polaris wasn't always the North Star and won't be in the future as the earth moves against the backdrop of stars and galaxies, that are also in motion. And Shakespeare referred to the North Star again when Julius Caesar proclaimed:

But I am constant as the Northern Star, of whose true fixed and resting quality there is no fellow in the firmament.

Because of the wobble in Earth's spin, the 'northness 'of Polaris is transient. In the time of the Egyptian Pharaohs, Thuban was the North Star. Over time, as the axis tipped, Thuban moved away, and Polaris approached the pole. In Shakespeare's time, Polaris was farther from the pole than now, but still close enough to be considered the North Star. But Caesar, who ruled in the first century B.C., would not have seen Thuban or Polaris as a fixed North Star. However, there was always a northern star even if the actual star changes over time.

A Sextant.

A sextant is used to measure the angle between an astronomical object and the horizon. Presumably, Younghusband or his navigator on *the Thetis* knew how to use a sextant. When a sextant is used on a moving ship, the image of the horizon and the celestial object will move around in the field of view. A sextant doesn't need a completely steady aim, as it measures a relative angle.

Learning how to navigate accurately was difficult and complex as even a cursory reading of *Moore's Practical Navigatiori* shows. The twentieth edition of the guide was published in 1828 with updating and corrections made by Joseph Dessiou. It is extremely detailed and very complex requiring an understanding of mathematics. Younghusband is unlikely to have had the mathematical skills to use this guide.

In 1834, up to fifteen different brigs were regularly crossing from Limerick to Quebec, so how were these navigation skill learned by the captains and navigators? There is no record of a navigation school for sailors in Limerick in the early 1800s. Navigation skills must have been passed on to apprentices on board ship. It's possible that some sailors on these transatlantic voyages had spent time in either the Royal or Merchant Navies. Many thousands of men had been released from service by the Royal Navy after the ending of the Napoleonic War so there was a supply of experienced sailors available.

However, it is more likely that navigation on these rudimentary cargo ships was kept as simple as possible. In summary, Younghusband set a compass bearing, selected a line of latitude and kept sailing until he reached the Canadian coast. Dead reckoning told him how far along his line of latitude he had travelled. Sightings on the North Star and determining local noon helped keep *the Thetis* on the chosen line. But of course, variable winds and currents ensured that arriving at the correct destination thousands of miles away on the far side of the Atlantic was far from simple.

A quadrant was the earlier navigation instrument, called because the sector is a quarter of a circle.

But when it came to the sextant not everybody was a fan:

Curse thee, thou quadrant!" Ahab then throws the quadrant on the deck, exclaiming, "No longer will I guide my earthly way with thee ... thus 6 trample on thee!"
– Captain Ahab of *the Pequod* in *Moby Dick*.

On the other hand, David Barrie on board the *Saecwen* in 1973 was very much a sextant enthusiast:

Taking the sextant away from my eye I looked at the scale and read off the angle... this was the sun's meridian altitude or mer alt. Colin took the sacred instrument from me and confirmed the reading. I looked up the sun's declination in the Nautical Almanac and made a few corrections to the observed angle. In a few minutes, to my astonished delight I completed the simple addition and subtraction sums that yielded our latitude.

Joseph Younghusband in 1834 must surely have been capable of using the sextant as competently as David Barrie in 1976.

Another basic aid was the lead and line which was used regularly to determine the depth of the water when anchoring but also this information could be compared to a contour line on a chart thus providing the location. An example of this method is the 100 fathom line which used by mariners entering the British Channel. The lead and line was also used to sample the sea bed.

While *the Thetis* was making her voyages across the Atlantic in the 1830s, an Irishman, Francis Beaufort was having a major influence on the provision of nautical charts around the coasts of Britain and Ireland. He also devised the first wind speed scale, the eponymous Beaufort Scale.

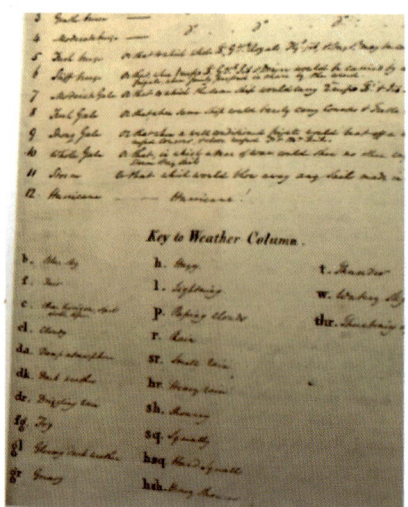

Beaufort's original wind speed scale.

Francis Beaufort was born in Navan, Co. Meath in Ireland on 27 May 1774. In 1829, Beaufort was elected as a Fellow of the Royal Astronomical Society. In the same year, at 55, Beaufort was appointed British Admiralty Hydrographer of the Navy, a post he held for 26 years. The scope of surveying was greatly improved under his tenure, both in home waters and overseas. But arguably his greater achievement was the provision of new nautical charts. When he was fifteen years old he was shipwrecked, which was apparently caused by inaccurate charts. This episode influenced his later career in the Navy. The production of new charts under his tenure increased from nineteen in 1830 to over twelve hundred in 1855 a major step forward in improving safety at sea.

Hydrographer of the Navy, Francis Beaufort.

Younghusband on board *the Thetis* had charts of the Gulf and the St Lawrence River. From 1759 to 1767, James Cooke, better known for his voyages in the Pacific, produced a comprehensive set of charts for this part of the Canadian Coast. He was involved in placing channel marker buoys on the St. Lawrence River from the Gulf of St. Lawrence to Quebec City. The first part of his survey was carried out during the Seven Years War with France and often within range of the French guns. Cooke's charts of The Gulf were still in use a hundred years later. Captain Henry Wolsey Bayfield was conducting a major survey of the St. Lawrence waterways during the 1830s but it's unlikely that his charts were available to Younghusband in 1834.

In the early 1830s, as *the Thetis* was crossing the Atlantic to Quebec, another ship left London and travelled to the Arctic. The ship, a converted ferry boat, was *the Victory,* a side-wheel steamer with paddles. It had been fitted

Henry Wolsey Bayfield (Surveyor).

with an engine however this was discarded early in the expedition. *The Victory* was captained by John Ross who intended to explore a northwest passage. Ross was also trying to redeem his reputation after cutting short a previous Arctic expedition because he had mistakenly identified clouds for a mountain range. He was accompanied on this voyage by his nephew James Clark Ross who on June 1st 1831, became the first European to reach the North Magnetic Pole.

James Clark Ross.

This portrait of James Clark Ross has the North Star in the top right corner of the picture.

In the 1820s, yet another Irishman was crossing the Atlantic. This man was heading not for Canada but was part of an expedition to the Arctic. Francis Crozier from Banbridge, Co. Down joined Captain William Parry's second Arctic expedition to explore the much sought after Northwest Passage. Francis Crozier became a close friend of James Clark Ross and joined him as second-in-command of HMS Cove in 1835, to assist in the search for twelve lost British whaling ships in the Arctic.

21 Dec 1835 Commissioned at Hull by Captain James Clark Ross to sail to Greenland waters to search for eleven missing whalers : the William Torr, and Lady Jane, both of Hull ; Viewforth of Kirkcaldy; Middleton of Aberdeen, were frozen in the land ice off the West Coast of Baffin's Bay, in about latitude 67ï¿ ½ N.; and the rest, including the Duncombe, Abram, Harmony, and Dordon, all of Hull; Norfolk of Berwick; Greenville Bay, Newcastle; Lady Jane, Newcastle, were reported to be drifting in the pack ice.
– Naval Data Base.

In 1845, Francis Crozier was second-in-command to Sir John Franklin on another expedition beyond the Arctic Circle. Franklin captained *the Erebus* while Crozier took command of *the Terror*. The expedition ended in 1848 with the loss of all 129 crew members including Crozier himself. By the middle of the 19th century, three Irishmen, Halpin, Beaufort and Crozier had made major contributions to exploration and safety in the North Atlantic.

Francis Crozier 1796-1848.

Knowing how to navigate at sea was important but knowing how to handle the complicated rigging of sailing ships was as critical. Many guides were available one of which was:

In Great Britain the naval arts are indigenous, and flourish with a superiority, which is the result of a vast demand for their various labours. But,

HMS Erebus.

singular though it is, the British Nation cannot boast of having taught or considerably improved them by the efforts of her press. Whatever may have been the cause of this does not at present much import; although curiosity would excite us to investigate, why these subjects have more engaged the attention of French authors.
– The Elements and Practice of Rigging And Seamanship, (1794), David Steel.

A rare piece of humility from David Steel acknowledging the French contribution. Richard Dana on board *the Pilgrim* describes how tacking or turning through the wind was carried out by the crew on a sailing ship like *the Thetis*.

The Chief Mate commanded on the forecastle and had charge of the head sails and the forward part of the ship. Two of the best men in the ship .... worked the forecastle. The third mate commanded in the waist and with the carpenter and one man worked the main jack and bowline: the cook the fore sheet, the steward the main. The second mate had charge of the after yards and let go the lee fore and main brace: three other light hands at the lee: one boy at the spanker sheet and guy: a man and boy at the maintopsail top gallant and royal braces and as all the rest of the crew tallied onto the main brace.
– Two Years Before the Mast.

This complicated sequence of events was impossible with large groups of passengers on deck. Sailing through the two estuaries of the Shannon and the St. Lawrence often required multiple tacks in close proximity to land. This meant passengers on board *the Thetis* were confined every day below decks for long periods.

Furling sails meant wrapping or unwrapping the large canvas sails around the yards, the horizontal timber beams. The crew climbed into the riggings in all weathers using rope ladders known as rat lines, above the deck. The large square sails required numerous crew members to get control of the flapping canvas as the helmsman pointed the ship into the wind to depower the sails. The crew had to be alert in hauling the sails as a large canvas sail can

Sailors unfurling a sail from a yard.

cause serious damage to the crew and ship if not brought under control as quickly as possible. Safety lines and life jackets were a thing of the future. There were two masts and up to twelve sails on *the Thetis* so this was not work for the faint hearted lacking a head for heights.

Another manual used on board sailing ships was *The Historic Naval Ships Association's Textbook of Seamanship*.

Aloft Topmen!
Take One Reef! Lay In! Down Booms!
Lay Down From Aloft!

Man the topsail halliards! Let go and overhaul the rigging! Clear away the buntlines, clewlines, and reef-tackles, and have them lighted up. Tend the braces! Let go the lee ones, and stand by to slack the weather ones. Set taut! HOIST AWAY THE TOPSAILS! Belay the topsail halliards! Trim the yards, Steady out the bowlines! and pipe down.

These commands were used on disciplined Navy ships but were they always in common usage on ships like *the Thetis*? *Aloft topmen* was an instruction to the crew to climb the ratlines into the rigging. The rigging system was so intricate that teamwork was essential. As these 'topmen' were 25m above the deck in all kinds of weather conditions clear instructions were vital. In one of Philip K. Allan's series of maritime novels *The Lea Shore*, his main character, Lieutenant Alex Clay readily admits to having no head for heights and dreads having to climb the rattlines. As he makes his way down slowly and carefully, the able seamen slid at speed down the stays from the topsail yards to the deck below leaving Alex Clay behind them to inch down slowly and carefully.

Continuing the account related to George Greaney in 1898, grandson of Dick Greaney who was an apprentice seaman on *the Thetis* in 1834.

The older crew used to tell us young ones to hold on very tight when aloft cos if'n we fell overboard they weren't going to stop to look for us as there were plenty more like us where we came from. Never was sure about that. Anyway always tied myself in as best I could. Even though I say so meself I was good enough aloft. Of course as I told you I liked it up there.

There are no photographs of *the Thetis* but one man created wonderful paintings of a similar brig. When I first saw these paintings of *the Beagle*, I thought they were painted at the time of Darwin's voyage to the Galapagos. But they were painted in 1970 by John Chancellor, an English artist.

Chancellor was a remarkable maritime artist who was renowned for the extensive research he carried out before undertaking a painting. He was born to English parents in the Lisbon area of Portugal in 1925, and at seventeen he joined the Merchant Navy where he was twice torpedoed. The painting by Chancellor of *the Beagle* is photographic in detail. The ship is sailing downwind along a coastline with a small number of sails deployed. Unusually, for this type of picture,

HMS Beagle by John Chancellor (1925-1987).

the sails are grey and not the usual pristine white in colour. Close examination of the painting shows fourteen sailors at various tasks around the ship. Anyone standing on Beale Strand in 1834 must have seen ships like this passing on their way to Canada. Julian Thomas of Art Marine had this to say about John Chancellor:

HMS Beagle by John Chancellor (close up).

John Chancellor was not a typical sea and ships man - he shunned the cliché subjects like mudscapes, clipper ships under full sail and the Battle of Trafalgar. He poured hundreds of hours into researching sea situations which interested him and took a particular joy in chronicling what he termed 'the unsung heroes', the small unglamorous barges, brigs and schooners.

Pictured on the previous page is a close up from the same painting which shows six seamen on the top yard unfurling the sail. A man on the right is sitting on top of the yard to get maximum leverage on the sail. Two others are standing on intermediate platforms supervising the unfurling of the mainsail. Another sailor is climbing a rattline to the front sail. And there are ropes everywhere. The bowsprit, the long timber pole at the front, has a number of lines providing support for the masts. This would seem to be a critical weakness in these sailing ships. If the ship crashed into a cliff, the bowsprit would take the full impact and the stays supporting the masts would be lost. The masts were then in danger of collapsing.

Another of John Chancellor's painting shows *the Beagle* under 'bare poles' in a mountainous sea in the Southern Ocean.

John Chancellor's the Beagle.

Imagine *the Thetis* in similar conditions, mid Atlantic with 217 immigrants crammed together and locked in the hold of the ship. In Chancellor's paintings, he goes to extraordinary lengths to get every detail correct. There isn't a rope missing. The stays holding the masts, halyards for raising the sails, guys, sheets and ratlines are all ropes made from the same material – hemp.. They are all beautifully recorded in John Chancellor's paintings.

Ropes on sailing ships were hauled by the crew in a coordinated fashion. A mistimed manoeuvre could cause serious problems. It was important on a ship like *the Thetis* to have one voice issuing commands and to have the crew responding immediately, without question. A way of achieving teamwork on board sailing ships of the era was to use sea shanties. Shanties were sung as the sails were hauled up or capstans were being turned. The shanty verses were sung by the team leader or shanty man with the rest of the crew joining in for the chorus. Shanties were call and response songs. Raising the anchor or pulling the ship alongside the pier on *the Thetis* needed a team effort so the shanty provided a rhythm to ensure the men worked in a coordinated way. The shanties fell into different categories depending on the task at hand. Some of the sea shanties are well known and have been recorded many times.

What will we do with a drunken sailor?
What will we do with a drunken sailor?
What will we do with a drunken sailor?
Early in the morning!

These songs were only sung on board ship. Sailors refused to sing them on shore regarding them as work songs only. The original lyrics of 'A-rovin'' were so bawdy, they were unfit to be sung in public.

In Amsterdam there lived a maid,
Mark well what I do say,
In Amsterdam there lived a maid,
Who was always pinching sailors' pay.
I'll go no more a-roving with you, fair maid.

A-roving, a-roving, since roving's been my ru-i-in,
I'll go no more a-roving with you, fair maid.

Each shanty was attached to a specific task on board. An anchor shanty would never be used for hauling sails.

Blow the Man down Oh blow the man down, bullies,
blow the man down
Way aye blow the man down
Oh blow the man down, bullies, blow him away
Give me some time to blow the man down.

The *Syracuse Daily Courier,* July 1867, quoted a lyric from the song, which was said to be used for hauling halyards on a steamship bound from New York to Glasgow.

Richard Weston was a passenger on *the John Dennison* which sailed from Scotland to America. He recorded the scene as they left the port:

And when everything was got ready a sailor sung the following words to a lively air and keeping time to the music as they all pulled.

"Pull away my hearty boys – pull away so cheerily
She moves along my boys-pull away so heartily.
We are all for America the wind is whistling cheerily
Then bouse away together boys and see you do it merrily."

Scottish postman Nathan Evans reached number one in the UK charts in 2020 with his version of a 19th Century sea shanty.

There once was a ship that put to sea
The name of the ship was the Billy of Tea
The winds blew up, her bow dipped down
O blow, my bully boys, blow (huh)

Soon may the Wellerman come
To bring us sugar and tea and rum (hey)
One day, when the tonguin' is done
We'll take our leave and go.

Take our leave and go.

'Wellerman' was thought to have been written by New Zealand whalers about an employee of the Weller Brothers shipping company in the 1830s.

The Beach Boys' 'Sloop John B' is also a shanty.

So hoist up the John B's sail
See how the mainsail sets
Call for the captain ashore
Let me go home
Let me go home
I wanna go home

"Under the Wave off Kanagawa", also known as The Great Wave is a woodblock print by the Japanese artist Hokusai was published between 1829 and 1833.

SIX
Crossing the Atlantic Ocean and the Gulf of St Lawrence

Thousands are sailing
Across the western ocean
To a land of opportunity
That some of them will never see
– The Pogues

A brig sailing in moderate seas.

Storms feature in accounts and paintings of transatlantic crossings in the 1800s, often for dramatic effect. However, as most of the crossings from Ireland in 1830-1834 took place across the summer months, storms were relatively infrequent on the voyage. The mid ocean part of the journey took about 30 days, the remainder being spent in the two estuaries. Summer storms last three to four days, so the weather for the rest of the voyage could be tolerated. But when there was a storm, conditions in steerage were appalling.

The misery reaches a climax when a gale rages for two or three nights and days, so that everyone believes that the ship will go to the bottom with all human beings on board.

Want of provisions, hunger, thirst, frost, heat, dampness, anxiety, want, afflictions, and lamentations ...

On the other hand, calm weather was frustrating for Younghusband on board *the Thetis*, because when there was no wind no progress was being made. Younghusband had a schedule to keep so sitting becalmed for days on end with hundreds of restless passengers was difficult to manage. Younghusband had to get to Quebec City, disembark 217 people, and organise to take out the berths in order to make room for the return cargo.

However, lack of wind did not neccesarily mean a calm sea. There can often be rolling swells tossing *the Thetis* about with little wind to control the ship, creating unpleasant conditions below decks for the passengers. What the captain and crew regarded as moderate conditions and ideal sailing weather might feel like a storm for the 217 people locked below deck on their first voyage at sea.

. . . during the voyage there is on board these ships terrible misery, stench, fumes, horror, vomiting, many kinds of seasickness, fever, dysentery, headache, heat, constipation, boils, scurvy, cancer, mouth rot, and the like, all of which

come from the old and sharply-salted food and meat, also from very bad and foul water, so that many die miserably.

0	0–1	0–1	0–1	Calm	–
1	1–3	1–3	1–5	Light air	Ripples with the appearance of scales are formed, but without foam crests.
2	4–7	4–6	6–11	Light breeze	Small wavelets, still short, but more pronounced. Crests have a glassy appearance.
3	8–12	7–10	12–19	Gentle breeze	Large wavelets. Crests begin to break. Foam of glassy appearance. Perhaps scattered.
4	13–18	11–16	20–28	Moderate breeze	Small waves, becoming larger; fairly frequent white horses.
5	19–24	17–21	29–38	Fresh breeze	Moderate waves, taking a more pronounced, longer form; many white horses are formed. Chance of some spray.
6	25–31	22–27	39–49	Strong breeze	Large waves begin to form; the white foam crests are more extensive everywhere. Probably some spray.
7	32–38	28–33	50–61	Near gale	Sea heaps up and white foam from breaking waves begins to be blown in streaks along the direction of the wind.
8	39–46	34–40	62–74	Gale	Moderately high waves of greater length; edges of crests begin to break into spindrift. The foam is blown in well-marked streaks.
9	47–54	41–47	75–88	Severe gale	High waves. Dense streaks of foam along the direction of the wind. Crests of waves begin to topple, tumble and roll over.
10	55–63	48–55	89–102	Storm	Very high waves with long overhanging crests. The resulting foam, in great patches, is blown in dense white streaks along the direction of the wind. The whole surface of the sea takes on a white appearance. The "tumbling" of the sea becomes more immense and shock-like. Visibility affected.
11	64–72	56–63	103–117	Violent storm	Exceptionally high waves (small and medium-size ships might be, for a time, lost to view behind the waves). The surface is covered with long white patches of foam lying along the direction of the wind. Everywhere, the edges of the wave crests are being blown into froth. Visibility affected.
12	73–83	64–71	118–133	Hurricane	The air is filled with foam and spray. Sea

Beaufort Scale.

Moderate sea conditions are described as having swells from 1.25 m to 2.5 m. Those conditions would have been of little concern to Younghusband and the crew. The problem with the word 'moderate' is that it's used in the Beaufort scale force four to describe the breeze. It appears again in the description of gale force five to describe the waves. But the word moderate is also used as a description of the wind in gale force seven. It is used to describe wave heights of over seven m in gale force eight. Gale force eight has wind speeds up to 40 knots. In gale force seven and upwards, Younghusband had to reef the sails as tight as possible and to have all hands on deck. In these conditions, passengers were confined to the dark and uncomfortable steerage where they were tossed about for hours or even days on end. The helmsman could anticipate the crest of the wave and bring *the Thetis* down the other side mitigating the stress on the boat. But more than 200 people confined in the hold of the ship on their first ocean voyage had no way of knowing when the next swell was coming. In those sea conditions, there was a constant state of confusion and fear below deck.

The Thetis in any condition above gale force five rolled from side to side and plunged bow first into the waves. The passengers were thrown about the hold. People were battered and bruised, and some had broken bones such was the force of the sea. There was a commotion with people crying, screaming, and praying loudly while trying to hold onto to their children and belongings.

The part of the voyage across the Atlantic Ocean typically lasted a month in the summer sailing season. There were exceptions when a ship could spend two months negotiating the Atlantic because of unfavourable winds. Trips of this length put intolerable strain on provisions and critically on fresh water. From force five upwards, the noise of the wind through the rigging and the waves breaking over the sides of the ship added to the general sense of panic. The passengers knew of ships that had recently been lost at sea. When the sea calmed, people tried to recover their possessions but inevitably tempers frayed and fights would break out adding to the discomfort and unease below decks. 200 people who had no sleep or food for days on end and were badly bruised by the rough conditions were a nightmare for the crew to manage and control. Unfortunately, some crews were ill disciplined and even at times violent towards the passengers.

On the worst vessels women were subjected to molestation and assault, and one guidebook observed with some heat that it was high time a determined effort was made "To prevent the violation of innocent and unprotected females by the brutes who are to be found sailing as officers or crews."
– The Great Migration.

Uncomfortable and unpleasant as this was, there were more dangers to come. Many tragedies and shipwrecks on these voyages occurred within sight of land, either in the Shannon Estuary or more often in the Gulf of St. Lawrence. Even though the charts of both estuaries were relatively accurate at the time, pinpointing many underwater dangers. The immigrants must have been delighted at the first sight of the New World. What they didn't realise was that the most dangerous part of the journey lay ahead.

The experienced crew members on *the Thetis* would know they were nearing land long before the first sighting. Birds were an early indication that land was close. Kittiwakes and shearwaters were welcome sights as they were signs that *the Thetis* was closing in on the Canadian coast. Changes in currents, floating debris, cloud patterns, and the colour of the water were indications to the experienced eye that land was near. And even smells were a sign of approaching land as some sailors said they could detect the scent of fires and cooking up to 100 km offshore. Having negotiated the Atlantic Ocean, *the Thetis* still had a 100 km obstacle course to endure before reaching Quebec City.

The first obstacle facing the crew was the danger from icebergs even though it was summer when most voyages took place. This area of sea from Greenland to the Gulf of St. Lawrence is known as Iceberg Alley.

From April to August every year, icebergs are visible from land along the eastern coast of Newfoundland. They can come in all shapes and sizes and colours from white to aquamarine. A collision with an iceberg would be catastrophic for small wooden ships like *the Thetis*. Even small icebergs or growlers as they were known could seriously damage the hull of a wooden ship. *The Titanic* hit one of the larger icebergs off the Canadian coast in 1912.

Richard Dana in *Two Years Before the Mast* sailing on *the Alert* describes the dangers of sailing through an ice field in the Southern Ocean.

But the field ice floating in great quantities and covering the ocean for miles and miles, in pieces of every size, large flat and broken cakes... This is very

difficult to steer clear of. A constant lookout was necessary for many of these pieces were large enough to have knocked a hole in the ship and that would be the end of us.
– Two Years Before the Mast.

A report of conditions in the Gulf of St. Lawrence in one week in May 1834 stated that:

Owing to the prevalence of easterly winds, our harbour is again blocked up with ice so as to prevent the ingress and egress of any burthen. All communication with the outports of the Island are necessarily much impeded.

In that one week, five ships ran into trouble in the Gulf of St. Lawrence because of the ice.

May 7 "Rebecca" from London to Quebec was lost in the ice near the Green Bank, crew saved.

May 8 "Economy" from Liverpool to Harbour Grace at about 6 leagues distance from St.John's struck a piece of ice and severely damaged. Assisted by the "Grace".

May 8 The brig "Agenoria" foundered about 7 leagues off Cape Spear owing to the damage received in the ice – Newfoundlander.

May 13 Brig "Maria" struck a piece of ice about midway between Baccalieu and Trinity harbour and sunk in about 20 minutes. (from Hamburgh to Trinity).

May 15 Brigantine "Favourite" from London and Poole for St.John's encountered the ice near St.Shott's and was totally lost after drifting on the rocks near that place.

The Gulf of St. Lawrence is the outlet of the Great Lakes, Superior, Michigan, Huron, Erie, and Ontario. The lakes are connected to the Gulf through the St. Lawrence River and then to the Atlantic. The Gulf is 26,000 sq km in area with an average depth of 150 m. In January, every year ice forms along the north shore of the Gulf. The ice that formed earlier in the estuary begins to drift through the Gaspé Passage. By the end of January, the western half of the gulf is covered by ice.

Iceberg Alley

Several of the vessels arrived at this port, passed immense Ice bergs about the banks of Newfoundland; and we perceive that the ship Euphrates at New-York gives the following description of them. On the 27th, in lat. 42, 30, long. 59, fell in with islands of ice, and continued passing them till next morning–counted 27 large ones, and saw a great number of smaller bodies. Passed within a cable's length of five of the largest.
– 20th May 1823, taken from the shipslist website.

The first entrance to the Gulf of St. Lawrence, when approaching from the north is a narrow strait at Belle Isle. Navigation in this strait can be extremely hazardous with strong tidal currents, and variable weather conditions which include gales and fog. Most ships in the age of sail wouldn't risk entering the Gulf through the Strait of Belle Isle, preferring to continue south to the much wider entry at the Cabot Strait. The Cabot Strait is between Cape Breton and Newfoundland but taking this route added two days sailing across the Gulf.

The passengers must have sensed an air of tension among the crew as they neared the Cabot Strait. Perhaps they wondered why all hands were on deck at the time. Passengers were known to change into their best clothes when they saw land as they thought they would be landing in Canada imminently. However, they still had weeks to go before reaching Quebec City. Robert Whyte on board *the Ajax* on the 10th July 1848 records the passengers delight when they first spotted the Canadian coast. But they weren't to disembark in Quebec City until the 3rd August.

The approach to Newfoundland was not, however, pleasurable from the point of view of navigation. Many a vessel was wrecked off the rocky coast, and it was a time for cautious sailing, foghorns, and sounding lines :
– The Great Migration, Edwin, C. Guillet

The Thetis had successfully negotiated the threat of icebergs but now a second obstacle faced the captain and crew. The foggiest place in the world is the island of Newfoundland. The cold Labrador Current from the north and the warm Gulf Stream Current from the east create perfect conditions for thick fog to form almost every day. Sebastian Junger's book *The Perfect Storm* documents the loss of the trawler *Andrea Gail* and her crew. *The Andrea Gail* sank while battling a storm on the Grand Banks in October 1991. During that storm, waves of 30m (100') and wind speeds of 200 km/h (120 mph) were recorded making the 'perfect storm'. Modern fishing trawlers with powerful engines and advanced weather forecast equipment were unable to cope with those conditions. In 1834, Younghusband on board *the Thetis* had no way of knowing what weather conditions awaited him when he arrived off the coast of Canada. Somehow, perhaps through luck and, of course, no small feat of sailing expertise, *the Thetis* and many other similar sized ships made it through the Gulf of St Lawrence to Quebec.

This was not an isolated seaway in the 1830s. On the contrary, it was a heavily trafficked channel. In May 1834, as many as thirty ships a day passed through the Cabot Strait en-route to Quebec City. A similar number were making their way back to their home ports in Britain and Ireland. Ships from New York, Boston and as far away as Jamaica made regular passage through the Strait on their way to Quebec and Montreal.

We spoke to a wherry which was conveying cattle from Nova Scotia to Newfoundland and learned from the steersman the bearings of St Paul's Island. We shortly afterwards passed a large fleet coming from the gulf and in the

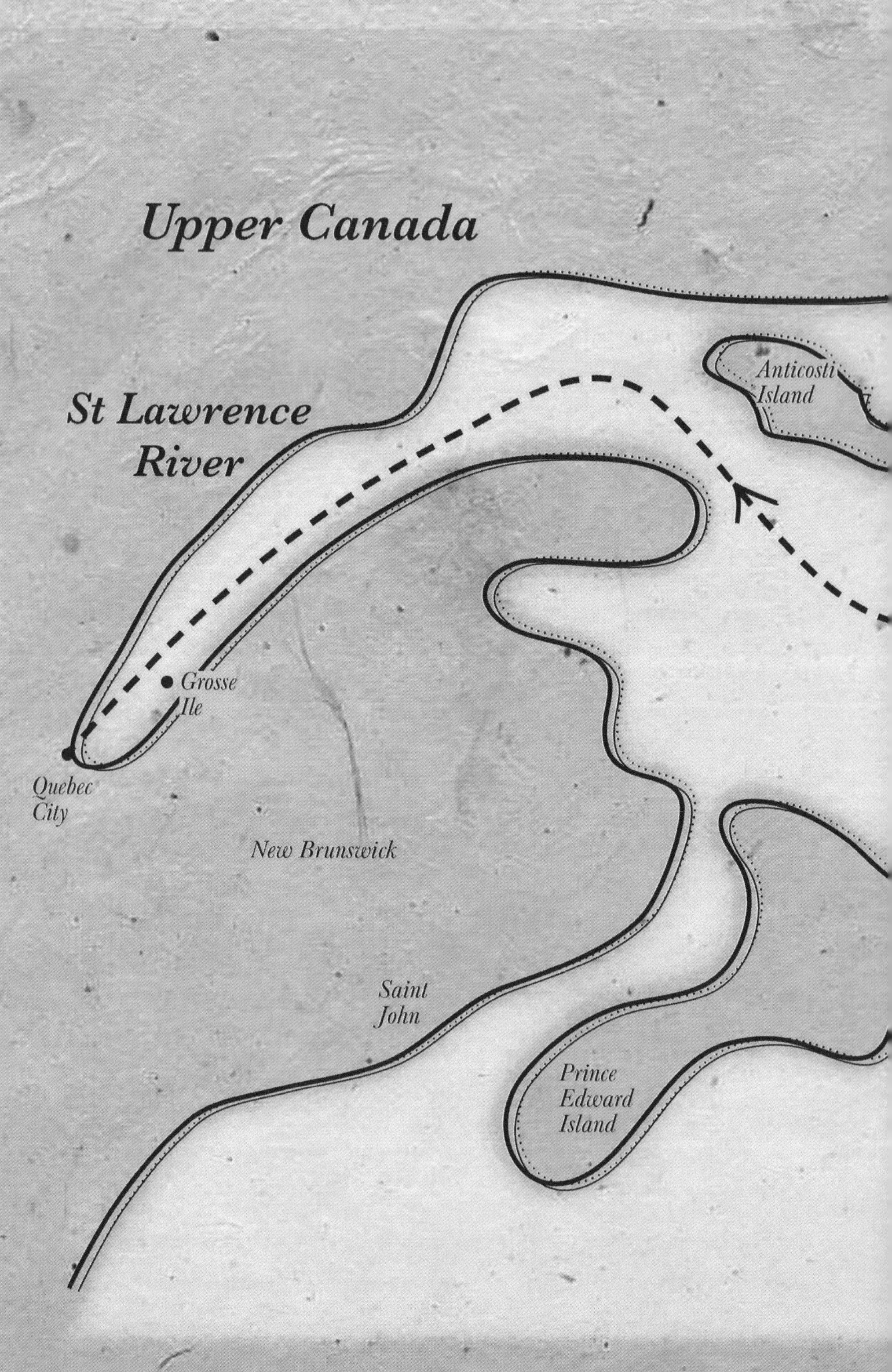

Upper Canada

St Lawrence River

Grosse Ile

Quebec City

Anticosti Island

New Brunswick

Saint John

Prince Edward Island

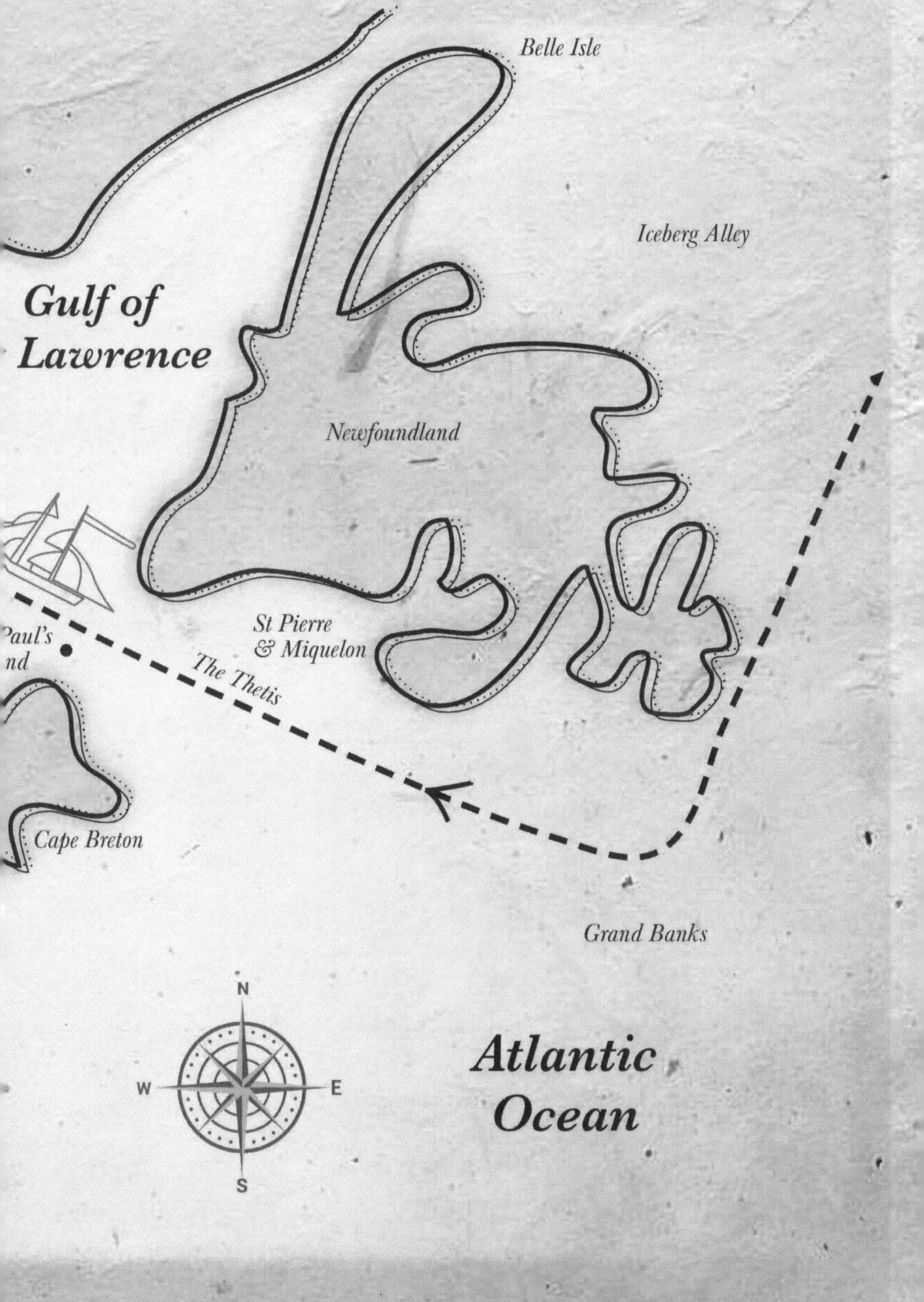

Belle Isle
Iceberg Alley
Gulf of
Lawrence
Newfoundland
St Pierre
& Miquelon
Paul's
nd
The Thetis
Cape Breton
Grand Banks
N
W
E
S
Atlantic
Ocean

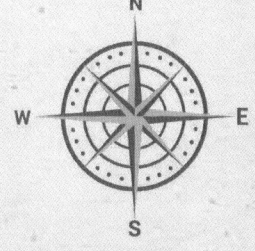

afternoon descried Cape North. The passengers expressed great delight at seeing land and were under the impression that they were near their destination, little knowing the extent of the gulf they had to pass and the great river to ascend.
– *Diary of a Famine Ship,* Robert Whyte.

As *the Thetis* passed through The Grand Banks, she met the Canadian fishing fleets working out of St. John's, just south of the Gulf of St. Lawrence. On board these fishing boats were hundreds of Irish seasonal workers from the Waterford area. These men spent the summers working on boats from St. John's before returning home to Ireland for the winter, explaining the strong familial links and accents between Waterford and this part of Canada.

And other fishermen were passing through the Cabot Strait. Fishing fleets from France used the French enclave of the islands of Saint Pierre and Miquelon as their base. As many as 200 French ships were recorded in the 1830s fishing on the Grand Banks. These ships crossed the Atlantic from the French ports of St.Malo, Fecamp, St.Brieuc and Dieppe. One report puts the number of French fishermen at over 8000 men every year. Being able to land on the French territory of Saint Pierre and Miquelon in the Gulf of St. Lawrence to dry and cure the fish was a great advantage for the French boats. The islands of Saint Pierre and Miquelon are now the last remaining French territory in North America.

At the opening to the Cabot Strait, *the Thetis* had to avoid the ice floes, fishing boats, nets, and passenger and cargo vessels coming and going from The Gulf of St. Lawrence. Herman Melville described passing through the Banks as a young sailor on his first voyage in his semi autobiographical novel *Redburn: His First Voyage (1849),* where he saw whales and a haunting shipwreck with weeks-dead sailors still on board.

The most strange and unheard-of noises came out of the fog at times: a vast sound of sighing and sobbing. What could it be? This would be followed by a spout, and a gush, and a cascading commotion, as if some fountain had suddenly jetted out of the ocean.

Seated on my Sampson-Post, I stared more and more, and suspended my duty as a sexton. But presently some one cried out–"There she blows! whales! whales close alongside!"
– Redburn: His First Voyage (1849).

In 1834, when *the Thetis* was on route to Quebec City, seventeen ships were shipwrecked in the Gulf of St Lawrence. 731 emigrants lost their lives. Later on, during a five-year period (1847-52) forty three emigrant ships out of a total of 6,877 sailings failed to reach their destination, which resulted in the deaths of 1,043 passengers. St. Paul's is a small uninhabited island situated 24 kilometres from the coast of Nova Scotia in the Cabot Strait. The island was nicknamed the 'Graveyard of the Gulf' as it is fog-bound throughout much of the navigation season, resulting in many shipwrecks.

It is a Newfoundland Fog; and we are yet crossing the Grand Banks, wrapt in a mist, that no London in the Novemberest November ever equalled. The chronometer pronounced it noon; but do you call this midnight or midday? So dense is the fog, that though we have a fair wind, we shorten sail for fear of accidents; and not only that, but here am I, poor Wellingborough, mounted aloft on a sort of belfry, the top of the "Sampson-Post," a lofty tower of timber, so called; and tolling the ship's bell, as if for a funeral.
– Redburn: His First Voyage (1849).

The incident, which should have made building a lighthouse on the island so urgent, was the wrecking of *the Jessie*.

The Jessie left Charlottetown, Prince Edward Island, on December 24, 1824. She was caught in a snowstorm which drove her onto St. Paul's Island. The passengers and crew survived on the island for eleven weeks but eventually died of starvation. Fires lit by the survivors could be seen from the mainland but because of the ice, no rescue vessel could reach the survivors. Advection sea fog is common in the Gulf especially in the early summer months. As the prevailing circulation shifts to the southwest, warm, moist air is pushed northward and cools from below. This causes fog to develop over the waters along the Atlantic coast. Unlike other types of sea fog, advection fog can often be accompanied by strong winds. Ships passing St. Paul's Island were often sailing blind, without warning lights or foghorns and because of the winds, were often travelling at their maximum speed of ten knots. On the contrary, having no wind in fog was equally precarious. Slowly drifting onto to the reefs because of lack of wind was a situation dreaded by captains of sailing ships. In his book *The Shoe and The Canoe* which he wrote about his journey in Canada between 1819 and 1827, Dr. John Jeremiah Bigsby describes the fog.

The fog was so penetrating as to soak with moisture the blankets in our state cabins and yet no one caught cold: so dense was it that sometimes we could not see the length of our small vessel.

The captain launched the rowing boat to throw sounding lines to the sea floor to help determine their position. The crew on the ship played bells and bugles so that the men on the rowing boat were able to return safely to the ship.

I shall never forget the vast magnifying effect of the mist on the ship ... She loomed into sight an immense white mass filling half the heavens.
– The Shoe and The Canoe.

If anchoring wasn't possible, the crew launched a row boat to try and tow the ship clear of danger but that was often not successful. George Vancouver who was one of the greatest maritime surveyors of the era, had exactly that problem on the west coast of America near Vancouver Island. His ship, *the Discovery* ran aground in August 1792 on hidden rocks in Ripple Passage when there was no wind. His other ship, *Chatham*, in the background in this picture, also ran aground on rocks the same day, about two miles away. Luckily both ships survived the ordeal with relatively little damage.

The Discovery on the Rocks in Queen Charlottes Sound, engraved by B.T. Pouney, c.1790-99.

George Vancouver's first captain was James Cooke on board *Discovery*. And earlier, when Cooke was captain of *the Endeavour*, he had a similar experience on the other side of the world in August 1770 when exploring the coast of Australia.

...at day break the vast foaming breakers were too plainly to be seen not a mile from us towards which we found the ship was carried by the waves surprisingly fast. We had at this time not an air of wind and the depth of water was unfathomable....
– *The Explorations of Captain James Cooke 1768-1771.*

But with extraordinary seamanship Cook and his crew managed to sail the *Endeavour* to safety through a small opening in the reef. On board *the Thetis*, Younghusband and his crew faced similar dangers as they passed St. Paul's Island. Too much wind or too little wind could cause similar disasters. Robert Fitzroy of *the Beagle* penned a vivid account of the wrecking of another ship called Thetis, a frigate of The Royal Navy. Fitzroy had begun his career on this ship but he wasn't on board when it was wrecked so this account is second hand. The last voyage of this Thetis began when it left Rio on December 4th, 1830 carrying $800,000 in gold bullion back to England.

Soon after eight one of the look-out men, named Robinson, said to another man on the forecastle, 'Look how fast that squall is coming' (this was the

cliff looming indistinctly through the rain and darkness), and next moment, 'Land a-head,' 'Hard a-port,' rung in the ears of the startled crew, and were echoed terribly by the crashing bowsprit, and thundering fall of the ponderous masts. The hull did not then strike the rocks, having answered the helm so fast as to be turning off shore when the bowsprit broke; but the lee yard-arm irons (boom-irons) actually struck fire from the rocky precipice as they grated harshly against it, the boom ends snapping off like icicles. All three masts fell aft and inward, strewing the deck with killed and wounded men.
– *Narrative of the Surveying Voyages of His Majesty's Ships.*

The 'bowsprit broke': The bowsprit is an extended timber pole sticking straight out from the bow. The stays from the masts are fixed to the end of the bowsprit which improves the angle of the stay. However, if a bowsprit broke the stability of the masts was critically reduced often causing them to collapse.

A bowsprit is a large boom or mast, cylindrically rounded, except at the outer end, which is square; it runs out over the stem, and stives or rises so as to make an angle of nearly thirty-six degrees above a horizontal line. Its principal use is to support the foremast by its stays, and carry sail to govern the fore part of the ship.
– *The Elements and Practice of Rigging And Seamanship,* 1794, by David Steel.

The original Lady Washington, or more commonly, Washington, was a historic sailing ship named after Martha Washington, and it sailed for about ten years in the 18th century.

This picture of the original *Lady Washington* under full sail shows how important the bowsprit was to the stability of the masts and rigging. The size of *the Lady Washington* was similar to *the Thetis*.

Unlike the Shannon Estuary, *the Thetis* didn't have lighthouses in the Gulf of St. Lawrence to rely on. Ships from Europe had been passing through the strait ever since the Italian John Cabot (Giovanni Caboto) sailed into the Gulf in 1497. But the Gulf had many hazards that ships had to avoid including Bird Rock and St. Paul's Island. The growing number of shipwrecks and the appalling loss of life eventually forced the government of Nova Scotia to act. In 1831, they placed provisions at a cove on St. Paul's Island southeast side to help shipwreck victims.

Before that, even when survivors managed to get ashore they were still in danger. When the Gulf froze over, the island was inaccessible so rescue was out of the question and many people starved to death as a result. But the lighthouse on St. Paul's Island was not built until 1838. It is hard to understand the inaction of the Quebec City port authorities in the early 1800s. The marine traffic from Europe passing through the Gulf had significantly increased year on year. The timber trade was extremely important for the economy of the region but every year there were serious incidents in the Gulf of St. Lawrence. Lighthouses might not have prevented all the accidents at sea but they would, surely, have reduced the fatalities.

In 1827, Edward Boxer, a Captain of the Royal Navy wrote the following to the Grand Admiral of Maritime Britain:

I have found a great need for lighthouses in the Gulf of Saint Lawrence. On this sea, navigation is so dangerous because of strong and irregular currents, and there is not a single lighthouse in all the Gulf. It is truly lamentable to find so many shipwrecks at different places on the coast... the number of lost lives is very large and certainly incalculable.

Admiral Henry Wolsey Bayfield from Hull in the U.K., pioneered hydrography in Canada. Even though he wasn't a formally trained surveyor, for forty years from 1816, he surveyed the Great Lakes, the St. Lawrence River and Gulf and the coasts of the Maritime Provinces. "I freely own that I am ambitious to complete the great labour which you have mentioned, extending from the head of Lake Superior to the western shores of Newfoundland," Commander Henry

Wolsey Bayfield wrote in May 1832 to Captain Francis Beaufort, Hydrographer of the British Admiralty. He was very aware on the dangers facing shipping in the Gulf. Bayfield wrote to the House of Commons committee recommending that there was a pressing need for lighthouses in the Gulf of St. Lawrence.

'From the entrance of the Gulf of St. Lawrence to Quebec is a very dangerous navigation yet there is but one lighthouse; a Floating Light should be placed at the dangerous pass called the Traverse and Lighthouses on Point des Monts and the East and West Points of the island of Anticosti.'
– Henry Wolsey Bayfield.

In another incident in August 1832, the *HMS Leonidas* left Halifax bound for Quebec with soldiers, passengers, general cargo and several tons of copper coins. While rounding the eastern tip of Cape Breton, the vessel struck Scatarie Island and sank. All aboard the *Leonidas* were rescued, but many others were not so fortunate when their ships struck the island. In 1834, as *the Thetis* travelled to Quebec, *the Fidelity* was wrecked on Scatarie Island and all 283 passengers and crew were drowned. In an article, entitled *The Wreck of the Astraea* published in the Dalhouse Review in 1950, D.C.Harvey detailed the dangers in The Gulf of St. Lawrence. In particular, he references the absence of lighthouses. *The Astraea* was wrecked in 1834 off Cape Breton with a huge loss of life.

'......in 1839 when the lights began to shine from St.Paul's Island and Scatari, both the number of wrecks and the loss of life in those that occurred diminished rapidly...'

The same article records that the British Government in 1836, offered to erect lighthouses on St.Paul's and Scatari if Lower Canada and the Maritime Provinces agreed to cooperate in defraying the expense of maintenance and administration. Yet still in 1841, hundreds of people lost their lives off the islands of St. Paul's and Scatari. Perhaps, they should have asked George Halpin to bring his Irish experience to build the lighthouses in The Gulf of St. Lawrence..

St. Paul's Island lighthouse.

There may have been another hidden danger in the Gulf of St. Lawrence. 'Pilotage', the practice of plotting a course in advance, is a vital part of marine safety. Every sailing course to this day stresses the importance of pilotage and passage planning. GPS chart plotters can lose power or signal, so an understanding of pilotage and charts is vital. This means picking a point on the chart to aim for, known as a way point. A line is drawn on a chart from the starting position to the way point which is the course the ship intends to take. A sailing ship can't stay on this straight line if wind and currents aren't favourable, but it is still the line the ship should return to.

A heading taken from the chart, Cape Breton to Anticosti Island reads 320°. This is a true north reading to distinguish it from magnetic north, the reading on the compass. In the Shannon Estuary in 2022, the difference between the two readings was 3° to 5°. But the difference between true north and magnetic north in the Gulf of St. Lawrence in 1834 was 28°. The chart reading of 320° had to be adjusted by 28° degrees to 348° when using the compass. If a ship's captain failed to apply the difference or subtracted instead of adding to the reading on the chart, then the ship was going to be seriously off course. In heavy fog that would prove to be a fatal error. We are reminded again of the absence of any formal training for the captains and crews of these ships. Another ship had a similar incident to the *Astraea* at St. Paul's Island.

For three days in the fall of 1834, dense fog made it impossible to shoot the sun at midday. The navigator of the ship. The Margaret, bound for Canada from Europe, deduced by dead reckoning that they'd passed Cape Race on Newfoundland. Three hundred nervous immigrants from Ireland were surely ready to put the infamous Cabot Strait behind them. Just a day later, Margaret slammed into the cliffs of St. Paul Island, 14 miles north of Cape North on Cape Breton Island. Miraculously, all of the passengers and crew made it ashore alive.

Maybe *The Margaret* didn't slam into the cliffs but slowly drifted onto the rocks when the wind dropped leaving the crew with no control. To Younghusband's credit, he brought *the Thetis* safely through the Gulf to Quebec City on at least four occasions. Many other captains were not so fortunate. Human error is another issue attributed to these shipping mishaps. Poor decision making or misjudgement in critical situations is often the real cause of sailing accidents. Younghusband was only twenty five years old when he first took charge of *the Thetis*. Timothy and Daniel Gorman were also in their twenties when they were appointed as captains. None of these seamen had any training in a maritime school. These young men sailed across the Atlantic in ships unfit for purpose with hundreds of immigrants on board. It is no wonder there were so many tragedies.

Jack London's short story *A True Tale Retold* about the wreck of the *Francis Spaight* makes a point about inexperienced sailors causing disasters at sea. However, his story lacks credibility because the captain of the ill-fated *Francis Spaight* was Timothy Gorman. Gorman was perhaps the most experienced seaman of his generation on the North Atlantic and was certainly not afraid of huge seas.

The Francis Spaight was running before it solely under a mizzentopsail, when the thing happened. It was not due to carelessness so much as to the lack of discipline of the crew and to the fact that they were indifferent seamen at best. The man at the wheel in particular, a Limerick man, had had no experience with salt water beyond that of rafting timber on the Shannon between the Quebec vessels and the shore. He was afraid of the huge seas that rose out of the murk astern and bore down upon him, and he was more given to cowering away from their threatened impact than he was to meeting their blows with the wheel and checking the ship's rush to broach to.

It was three in the morning when his unseamanlike conduct precipitated the catastrophe. At sight of a sea far larger than its fellows, he crouched down, releasing his hands from the spokes
– A True Tale Retold, Jack London.

There's no record of anybody 'rafting timber' on the Shannon. This is yet another example of how these seamen were regarded at the time. Jack London's account of the *Francis Spaight* is very far from a true tale retold.

The crew were 'indifferent seamen at best'? Perhaps they were, but they were also brave or perhaps even reckless. However, it was recognised that there was a problem with the training of sailors. Nothing was done to address the issue until years later when Spaight's son, James was involved in improving seamen's conditions and training. In the meantime, hundreds of immigrants lost their lives in the Gulf of St. Lawrence due to unsuitable ships and lack of guiding lights. But it was much easier to blame the captains and crews rather than facing up to the real issues of overcrowding, poor ships and lack of investment in safety at sea.

On a typical forty-day voyage, only two days were spent in the Shannon Estuary, but crossing the Gulf of St. Lawrence was a much longer journey. The Gulf measures 500 km from the Atlantic to Anticosti Island at the head of the St. Lawrence River. From there to Quebec City is another 320 km. The journey from the mouth of the Gulf to Quebec City could take a least a week to negotiate.

The passengers on *the Thetis* saw strange sights as they passed through the Gulf. There were dolphins in the Shannon but there were Blue Whales and Beluga Whales in the Gulf. Blue Whales are the largest animals on earth and they arrive at the Gulf of St. Lawrence in March. About 1,000 white Beluga Whales live in the Gulf all year-round and they must have been an intriguing sight for the passengers. There are many accounts of passengers on board the emigrant ships fishing for mackerel with makeshift hooks and line as they arrived at the mouth of the Gulf. Catching fresh fish made a welcome change of diet from the undercooked, stale, and rancid provisions that were left on-board ships like *the Thetis*.

The Thetis having left the danger of St. Paul's Island behind, headed in a northwest direction towards Anticosti Island 200 km away at the head of the St. Lawrence River. Anticosti Island has been called the strangest place in Canada. It was the next obstacle to be dealt with by *the Thetis*. St. Paul's Island was often called the Graveyard of The Gulf, but Anticosti Island had a similar reputation with over 400 shipwrecks recorded in the vicinity. As Guillet describes:

Even more dangerous than St. Paul's Isle was Anticosti, a large island without bay or harbor, "the grave of many a tall ship". Like St. Paul's it had a lighthouse in the 1840's, but earlier there was merely a house at either end, equipped and provisioned to shelter the shipwrecked. Approaching it in a fog and drifting with the tide without the aid of a pilot, the Brunswick was in great danger of grounding upon it. Every captain, in fact, who found himself within ten miles of the island immediately sheered off on another tack, for it is surrounded by sunken reefs.
– *The Great Migration,* Guillet, Edwin, C.

As *the Thetis* passed by Anticosti, the island was uninhabited apart from a legendary character called Louis Oliver Gamache. He was a Frenchman who through a series of misadventures ended up on Anticosti with his family. He may have been either a hero or a villain. Gamache was applauded for helping victims of shipwrecks and then alternatively accused of deliberately luring ships onto the rocks and reefs around Anticosti. Gamache's story is well documented elsewhere. He often clashed with the authorities and there is even a mention of smuggling. It would have been a very elegant answer if there was evidence of Joseph Younghusband and Louis Oliver Gamache being co-conspirators in smuggling tobacco. Anticosti Island was a perfect location to load contraband tobacco on *the Thetis* on her home journeys. But no such evidence exists. There are many stories about *Anticosti Island*, including the story of the French Chocolate King, Henri Menier, who bought the island in 1895 for $125,000. He turned it into his private hunting reserve. Menier imported a herd of white-tailed deer for hunting purposes. There are now over 160,000 of these deer on the island. And they are still being hunted to this day.

Having negotiated the waters off Anticosti, Younghusband turned *the Thetis* northwest up the St. Lawrence River. In *The Backwoods of Canada,* Catharine Parr Strickland, a young English writer, described her voyage along the St. Lawrence:

The misty curtain is slowly drawn up, as if by invisible hand, and the wild, wooded mountains are partially revealed, with their bold rocky and sweeping bays. At other times the vapoury volume dividing, moves along the valleys and deep ravines, like lofty pillars of smoke, or hangs in snowy draperies among the dark forest pines.

A watercolour by Elizabeth Simcoe 1792 A bend in the St. Lawrence River.

Final Voyage of the Thetis

An idyllic scene which must have seemed wonderful to the immigrants on board but then *the Thetis* had a problem with another island.

> *... A vast fleet of vessels lying at anchor told that we had arrived at Grosse Isle; and after wending our way amongst isles and ships we dropped in the ground allotted for vessels upon arrival, and hoisted our ensign at the peak as a signal for the inspecting physician to board us.*
> – *The Great Migration*, Guillet, Edwin, C.

Grosse Île (Big Island) despite its name is a small island about eight km² in area, situated 46 km downstream from Quebec City. In 1832, the authorities were alarmed at the outbreak of disease, typhoid and cholera in the city so they took action to stem the spread of disease.

Mary McLean Walsh writing about her 1832 arrival in Quebec said:

Grosse Ile and the Irish Memorial National Historic Site.

> *On the tenth of May, 1832, my parents with their eight children sailed from Ireland to America; and, although the other passengers fared very well, and notwithstanding I was a perfectly healthy Irish girl of sixteen, during the whole voyage of one month, I was ill.... We were much relieved to be landed in Quebec but imagine our feelings at finding ourselves in a plague-stricken city; where men, women and children, smitten by cholera, dropped in the streets to die in agony.*
> – Historian Kerby Miller, from a memoir dictated by Mary McLean Walsh to her daughter Sarah Kirwan in 1890.

In February 1832, the Assembly of Lower Canada (Quebec) established a quarantine station on Grosse Île. The island was an ideal location to assess the well-being of the passengers and hopefully contain the cholera outbreak and continued to operate until 1937. David Cragg, a widower on board *The Six Sisters* packet ship from England arrived at Grosse Ile in June 1833. Cragg and his eight children were disembarked on the quarantine island.

...they appoint us a booth in a large building made like an entrance house of great length. All this time there is 1199 passengers performing penance here... we slept at night on beds made on the floor side by side, all together like a lot of pigs...the doctor in all pomposity came, looked us over, fooled among our goods for about two minutes and then said we were at liberty to go.
– Quoted by Barbra Dickson in *Quarantined, Grosse Ile.*

On that second last trip in 1834, *the Thetis* entered the St. Lawrence River with over 200 passengers. As there had been a cholera outbreak on board *the Thetis,* Younghusband flew a blue flag on arriving at the quarantine island. He wasn't allowed to continue to Quebec City until the doctors on Grosse Ile gave him the all-clear. The passengers who showed no symptoms had to remain on board *the Thetis* during this period.

It's not clear why the authorities used a blue flag as the 'yellow jack', a plain yellow banner, was historically used to signify a vessel had disease on board. The 'yellow jack' represents the letter Q or Quebec in maritime signal flags and ships continued to fly the flag until granted 'free pratique' by the port authorities. *The Thetis* as mentioned earlier had the Red Ensign flying from the stern, the most senior position on the ship. If the flag was 'defaced' by the harp, the authorities could immediately identify it as coming from Ireland.

Several people of all ages had died on the journey on *the Thetis* as she crossed the Atlantic. They were recorded as having died from common cholera. Just days after leaving Ireland, on the 18th April 1834, five year old Lawrence O'Brien died. William

O'Brien, aged ten, died on the 20th April. Biddy Kelly, a six month old baby, Michael Hurley aged two, Michael Kane also aged two, John O'Dowd (37), Judy Behan (18) all died between the 22nd and the 28th April on board *the Thetis* in the mid Atlantic. But the most poignant of all must be the story of Mary Kennelly. She died on board *the Thetis* from sea sickness. She was eighty years old. It is hard to understand why this old woman boarded *the Thetis* in Limerick to travel for seven weeks across the Atlantic in a dark, filthy and diseased ship's hold. Of the 217 passengers on board the Thetis in April 1834, the names of the six people who died on board are the only ones that have survived.

The Thetis didn't arrive in Grosse Ile until the 22nd May 1834. Knowing that cholera was rampant on board the ship, the five weeks of the journey from the middle of April must have been a terrifying experience for both passengers and crew. There was nowhere on board *the Thetis* to self-isolate from the threat of typhoid or cholera. Somehow, the crew had to be kept free from disease, as the lives of everybody on board *the Thetis* depended on the crew being fit and well.

In 1832, Catherine Parr Strickland's sister, Suzanna Moodie also wrote about her arrival at Grosse Ile in her book *Roughing it in the Bush*. She made derogatory comments about the people, mainly Irish, she encountered on the quarantine island.

Here we encountered a boat, just landing a fresh cargo of lively savages from the Emerald Isle. One fellow, of gigantic proportions, whose long, tattered great-coat just reached below the middle of his bare red legs, and, like charity, hid the defects of his other garments, or perhaps concealed his want of them, leaped upon the rocks, and flourishing aloft his shilelagh, bounded and capered like a wild goat from his native mountains. "Whurrah! my boys!" he cried, "Shure we'll all be jintlemen!"
– Roughing it in the Bush, Suzanna Moodie.

But for many, the journey to a new beginning ended almost immediately. Thousands died shortly after leaving Ireland, either on the overcrowded ships or in the squalid quarantine stations. In the spring of 1832, medical staff arrived on the island and set up the hospital. There were two basic shelters built, each of which held 300 people. The first cases of cholera were discovered in the spring of 1832. It took until November before medical authorities managed to contain the disease. But later, between 1840 and 1847, Grosse Ile was overwhelmed with immigrants fleeing the Great Famine in Ireland. Between 1832 and 1937, Grosse Île's term of operation, the official register lists 7,480 burials on the island. In 1847 alone, 5,424 burials took place, the majority of whom were Irish immigrants. In that same year, over 5,000 Irish people on ships bound for Canada are listed as having been buried at sea.

A journalist of the time described the erection of the memorial:

Peace has its victories, but it also has its tragedies. The huge Celtic Cross that majestically raises itself on high from its island foundation will serve to remind us that there are nobler heroes found in lowly places than in the dramatic din of the battlefield. It will serve to make known the heroic dedication of those brave people who risked their lives in coming to the rescue of these poor victims of a man-made calamity.

Over 160 years later, on the 1st October 1998 Mary McAleese, the then President of Ireland, visited Grosse Ile to remember the Irish who were buried there.

We cannot change the past - but we must not forget it. Let us ensure that the tragedy of Grosse Ile, the bravery of Grosse Ile and the humanity of Grosse Ile is never forgotten and never denied.
– President Mary McAleese.

Children of the Gael died in the thousands on this island having fled from the laws of the foreign tyrants and an artificial famine in the years 1847-48. God's loyal blessing upon them. Let this monument be a token to their name and honour from the Gaels of America. God Save Ireland.
– *Inscription on the Celtic Cross.*

After leaving Grosse Ile, *the Thetis* had a day's sailing up the St. Lawrence River before reaching Quebec City.

The 22nd of September, (1832) the anchor was weighed, and we bade a long farewell to Grosse Isle. As our vessel struck into mid-channel, I cast a last lingering look at the beautiful shores we were leaving. Cradled in the arms of the St. Lawrence, and basking in the bright rays of the morning sun, the island and its sister group looked like a second Eden just emerged from the waters of chaos........The air was pure and elastic ,the sun shone out with uncommon splendour, lighting up the changing woods with a rich mellow colouring, composed of a thousand brilliant and vivid dyes.
– *Roughing it in the Bush,* Suzanna Moody.

When *the Thetis* rounded the bend in the river at Isle d'Orleans, the passengers were greeted by a dramatic scene. There were often over a hundred ships of differing sizes and types moored in the river beneath the Citadel on Cap Diamante. There were ships from Great Britain and Ireland, and many others from New York, Boston and the Caribbean. Quebec City had a thriving ship building industry with over a hundred new ships being launched every year. The Royal Navy had a contingent of ships permanently based in Quebec City. In 1834, this was a very busy port with rafts of timber being loaded onto the same ships that had just recently discharged hundreds of immigrants. On arriving at Quebec City port, the passengers from *the Thetis* were immediately and without much ceremony, disembarked on the quay side.

"It is a curious and rather painful sight," wrote an eye-witness, "to watch the emptying of a newly-arrived cargo of emigrants on the unknown shore. Squalid, thinly clad, and far from clean, you instantly distinguish the bony Irishman with his wife and all the children, dragging an ill-packed bundle tied with a bit of rope, which is made long enough by the help of a strip of ticking or a list border. They slide their bundle - their all of worldly wealth - down a plank, and having drawn it aside on the dock, they hang helplessly around it, the children tumbling on it, till the ship has disgorged her motley company, and all are ready to appear at the Emigrant Office.
– The Great Migration, Guillet, Edwin, C.

For immigrants from France, and Belgium, the lower town of Quebec City must have seemed very familiar. Even now, walking through the narrow-cobbled streets of Old Quebec is like visiting the towns in Northern France. It's not just the architecture of the houses and the street layout that reminds one of France. French is the main language spoken here. Old Quebec, or, more correctly, Vieux-Quebec, is a warren of small narrow streets with houses dating back to the 17th century. Walking through this tourist area, one is struck by the fact that these are the same streets that the passengers from *the Thetis* walked in the 1830s after crossing the Atlantic. But strangely, there is

A streetscape in Old Quebec City (2023).

no monument or even a plaque on the quay side in Vieux-Quebec to remember the hundreds of thousands of immigrants from all over Europe who landed here in the first half of the nineteenth century.

On arriving in Quebec City, Younghusband had a tight schedule to adhere to, so the welfare of the immigrants was no longer his priority. *The Thetis* had to be readied for the return journey to Limerick. The stone ballast had to be taken from the hold and disposed of in the bay, or, as happened on occasion, sold on shore. The wooden passenger berths had to be removed, the hold cleaned and fumigated, and the timber loaded on board. Cleaning the steerage and the lower orlap deck was a much-hated task. There was six weeks of accumulated human waste and rubbish to be cleared out of *the Thetis*. And then, there was the matter of tobacco to be arranged. The passengers were left to fend for themselves. In 1834, many of them were well prepared for the next phase of their journey as this article from a Canadian publication points out.

Irish immigration is often presented as a tragic epic in which victims of famine were forced to flee their homeland. The truth is otherwise. It is a tale of how hope and hard work gave Canada its stalwart Irish population. Most of the Irish left of their own free will and financed their sea crossings themselves or were helped by family and friends to meet the cost. Far from being powerless

victims, they planned their departure carefully and were highly knowledgeable on the economic advantages which Canada offered.
– Atlantic Canada's Irish Immigrants: A Fish and Timber Story (The Irish in Canada), Campey, Lucille H. A.

Clearly, this was completely untrue during the Famine Years of 1840 to 1847. However, even in 1834, some weren't well prepared and perhaps the last word about Irish immigrants landing in Quebec City should be left to John Jeremiah Bigsby in *The Shoe and the Canoe.*

This misery does not touch the native poor but the fever stricken naked and friendless Irish -a people truly scattered and peeled -who year after year are thrown in shoals upon the wharfs of Quebec from ships which ought to be called "itinerant pest-houses". These unwelcome outcasts are crowded without proper provision into vessels fitted up almost slave-ship fashion by the agents of impoverished and unprincipled landlords...much of the guilt certainly lies with the Irish Government.
– The Shoe and The Canoe.

Which view is correct? A fever stricken, naked and friendless people or a people who were 'highly knowledgeable on the economic advantages Canada offered'. Perhaps, both perspectives have an element of truth to them.

This monument in Toronto remembers the Irish who arrived in Canada during the Famine years.

SEVEN
The Return Journeys July and October 1834

Cape Breton (46.25 N, 60.85 W) to Limerick (52.87 N, 10.56 W)

Ships loading timber in Quebec City (1860).

On the return journeys, *the Thetis* had only eighteen experienced sailors on board. It is possible that occasionally a small number of returning passengers were on the ship. It is inevitable that some of the younger immigrants or people with families back home would decide to return to Ireland. Benjamin Franklin was able to make eight trips to Europe from New York in the late 1700s so return journeys were possible. The lack of any cabin accommodation on *the Thetis* made this unlikely.

The hold of *the Thetis* was now full of timber, and, of course, tobacco. The smell of resin from the wood and the contraband tobacco on the ship was overpowering. But it was so much better than the smells created by hundreds of people crammed into the space beneath decks in the hold. The air of tension and danger was gone. Death and disease were no longer threatening the health of the crew. They were going home and were going to be well rewarded for their efforts. And they wouldn't have to cross the Atlantic for another six months.

If everything went well, with the prevailing winds behind them the Gulf Stream would push them home. *The Thetis* could take a direct route. But navigation would have to be much more precise. Ireland is a much smaller target than Canada. Crossing diagonally over lines of longitude and latitude required a greater degree of accuracy from the navigator. Of course, maybe they retraced their steps northwards until they reached the 52 ° line of latitude.

In 1768, while in England, Franklin heard a curious complaint from the Colonial Board of Customs: Why did it take British packets several weeks longer to reach New York from England than it took an average American merchant ship to reach Newport, Rhode Island, despite the merchant ships leaving from London while the packets left from

Benjamin Franklin.

Falmouth in Cornwall? Franklin asked Timothy Folger, a Nantucket Island whaling captain, for an answer. Folger explained that merchant ships routinely crossed the current -while the mail packet captains ran against it.
– Gulf Stream, Wikipedia

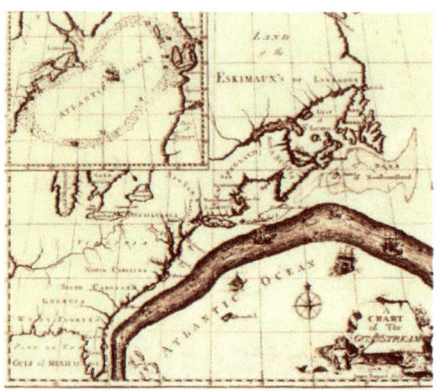
The Gulf Stream by Timothy Folger 1769.

Franklin had Timothy Folger sketch the path of the current on a chart of the Atlantic and add notes on how to avoid the current when sailing from England to America. Franklin then forwarded the chart to Anthony Todd, secretary of the British Post Office. Franklin's Gulf Stream chart was printed in 1769 in London, but it was mostly ignored by British sea captains.

Ships bound for Europe from Cape Breton all faced the same dilemma. Should they retrace the line of the incoming voyage and sail the northern parallel? Or should *the Thetis* sail the shorter route and plot a course directly for Loop Head, 2,500 miles to the northeast? If the captain was confident of his navigation skills, then he would use the power of the Gulf Stream and the prevailing southwesterlies to arrive at the home port a week earlier. But the skill needed to arrive at Loop Head, while relying on a sextant and dead reckoning while crossing lines of longitude, was considerable. Staying in the Gulf Stream could increase the speed of *the Thetis* by as much as three knots. If they had clear skies to take sightings on the moon and stars, then this was relatively straightforward for an experienced navigator. Perhaps, they took a safer third option and sailed the lower 46° parallel before turning north towards Ireland and home.

But on the return journey, there could be reminders of how dangerous this sea could be as Herman Melville recounts.

It was a dismantled, water-logged schooner, a most dismal sight, that must have been drifting about for several long weeks. The bulwarks were pretty much gone; and here and there the bare stanchions, or posts, were left standing, splitting in two the waves which broke clear over the deck, lying almost even with the sea. The foremast was snapt off less than four feet from its base; and the shattered and splintered remnant looked like the stump of a pine tree thrown over in the woods...

Lashed, and leaning over sideways against the taffrail, were three dark, green, grassy objects, that slowly swayed with every roll, but otherwise were motionless. I saw the captain's, glass directed toward them, and heard him say at last, "They must have been dead a long time." These were sailors, who long ago had lashed themselves to the taffrail for safety; but must have famished – Redburn His First Voyage, Herman Melville.

An experienced crew with a capable captain would not have been concerned about the weather on the return trip during the late summer and early autumn. If the cargo was well secured, *the Thetis* could have dealt with most Atlantic weather. But there was another difficulty to face. Overloading of these timber ships had become a major issue. Some of the shipping disasters in this era can be attributed to the practice of deck loading. Not only was the hold full of timber, but it was common practice to increase the cargo by putting extra timber on the open deck. This had the effect of destabilising the ship, particularly if the timber had not been correctly secured.

Deck loading caused the centre of gravity of the ship to be moved forward, disrupting the balance between the power in the sails and the ballast below. The extra load to the front of the ship caused the bow to dive under the waves, often with catastrophic results. The issue of deck loading the timber ships crossing the Atlantic was regularly debated in the House of Commons. If *the Thetis* was a flat decked brig without a foc'sle and was carrying timber close to the bow, then the danger of capsizing was greatly increased. In his Gothic novel, about a ghost ship called *the Flying Dutchman,* Frederick Marryat describes the problem very succinctly,

Another minute, and the vessel was swung round on her broadside to the sea, and lay on her beam ends.
– *The Phantom Ship,* Frederick Marryat.

'Beam ends' was to be avoided at all costs because the ship was tipped sideways and the balance between the sails and the weight of the keel was on a knife edge. Too far over and the ship could not right itself and would capsize. However, *the Thetis* had the southwesterlies behind her so fewer tacks were required and easier sailing resulted. Despite the overloading, the first return journey from Quebec in 1834 was uneventful. *The Thetis* loaded with timber and presumably, smuggled tobacco, was back in Limerick in mid-July. *The Thetis* might even have got close to the twenty one days boasted about earlier in the newspaper notice.

Perhaps because of the quarantine restrictions, Francis Spaight decided that the next trip from Limerick to Quebec would be without passengers. Being held in Grosse Ile for two weeks was a major inconvenience for ship owners. Any long delays at this time of the year and the ice in the Gulf would keep *the Thetis* locked in Quebec City until the following spring.

The Thetis set sail for Canada from Limerick, again, in August 1834, 'in ballast' only. After they arrived in Quebec City, Younghusband collected his cargo of timber and set sail almost immediately for Ireland. There's no record of the weather conditions that *the Thetis* faced on this last trip crossing the Atlantic, but according to the information on the Shiplist website, they made the trip in good time. Presumably, therefore they didn't face any adverse conditions.

The first sign of land, the lookout on *the Thetis* saw on arriving off the Irish Coast was the lighthouse fire on Loop Head. The fire was from a brazier tended by lighthouse keepers and was visible at least fifteen miles out to sea. As a young and relatively inexperienced captain, Younghusband was probably relieved that his navigation had brought *the Thetis* this far. All they had to do was reach Loop Head and the relative safety of the Shannon Estuary. *The Thetis* had made this part of the voyage many times so the crew knew calmer waters lay ahead.

If they were lucky, they would reach the mouth of the Estuary as the tide was turning. But was it just luck? Was Younghusband able to time the ocean crossing so he arrived at Loop Head as the tide was turning? He'd know the times of the tide as he left Ireland so he could calculate the time of the full tide when arriving back at Loop Head. With a favourable wind direction and an inward flowing tide, *the Thetis* could achieve a speed of 10 knots from Loop Head to Kilrush. If they were unlucky with the timing, *the Thetis* would have to wait off Loop Head for up to six hours while waiting for the tide to turn. An hour after entering the Estuary, the ship passed the second lighthouse fire at Kilcredan Head. They could also see the light on Scattery Island in the distance. But, they had another issue to deal with before reaching the Kilrush anchorage.

Bales of contraband tobacco had to be offloaded from *the Thetis* as soon as possible. Younghusband needed to take on a pilot at Kilrush, Cappa or Scattery depending on which pilot boat was first to arrive. There were pilots in Kilbaha but Younghusband neither needed nor wanted a pilot on board until Kilrush. It was difficult enough to have sixteen crew members sworn to secrecy about the contraband tobacco without running the risk of a pilot informing the Customs officials. The other risk was from the Revenue officials who could board *the Thetis* as she made her way up the river to Limerick. The remaining forts from the Napoleonic Wars were still manned by soldiers in 1834 who were keeping watch twenty four hours a day.

But smuggling of tobacco in the Shannon Estuary wasn't a risk-free enterprise. Trouble was always close at hand.

The deadly intent and violent nature of smuggling was revealed in reports such as that of an incident in 1821 off the County Clare coast in which a stranded smuggler's cargo was being transferred to another vessel. The

smugglers were then engaged by a local militia who were driven off with the loss of one life. The smugglers then set fire to the munitions store on the ship and escaped with their cargo and crew, an action which demonstrated their ability to tackle a local militia and to absorb the loss of a ship.
– The State's Attitude and Response to the Threat Posed by Tobacco Smuggling in Ireland 1780–1850, Irish Social and Economic History, Vol. 48, Dr. Sean Whitney.

When *the Thetis* arrived at Loop Head, after crossing the Atlantic, Younghusband had to alert the watchers on shore that they had arrived. He had to ensure that they would be ready to offload the contraband tobacco from the ship. The people on shore could only know when *the Thetis* arrived off Loop Head within a window of two to three weeks. One possibility is that when *the Thetis* came in sight of the lighthouse on Loop Head, the keeper recognised the company flag. He had to record the arrival of *the Thetis* so the lighthouse fees could be collected in Limerick from Francis Spaight.

The pilots in Kilbaha would know *the Thetis* was on her way up the Estuary so they could have alerted the watchers on shore. A dispatch rider might have been sent to relay the information to the waiting smugglers on the Kerry side of the Estuary. Or else by either a system of light signals from Loop Head to Kerry Head or from *the Thetis* to the Kerry shore would alert the people on the shore. A signal to watchers on Kerry Head near Ballyheigue would alert members of the local tobacco combine that *the Thetis* had arrived.

Younghusband was at the mercy of the tides and so couldn't choose to land the tobacco under the cover of darkness if the current was running against *the Thetis*. He had to offload the contraband in Beale an hour or so after passing Loop Head no matter what the time of the day or night. This is speculative but somehow or other, *the Thetis* was met in Beale by members of the tobacco combine. The Combine needed time to arrange men and transport to be at the beach to move the tobacco as quickly as possible to avoid being intercepted by the Revenue Officials.

About two hours after passing Loop Head, *the Thetis* ran aground on Beale Bar. We are told that nine crewmen were drowned when the ship became stuck on the reef. The remaining crew including the captain, were saved. Customs officials inspected the ship three days later and discovered the smuggled tobacco. The crew and Joseph Younghusband, who was waiting in Tarbert, were arrested by the Custom officials. They were marched to Tralee and tried. Or so the story goes. But none of that makes much sense. The facts don't fit the official story, particularly the part where nine sailors were washed overboard and lost. The *Kerry Evening Post* of the 3rd December 1834 reported that:

It then blowing a gale at N.W. and afflicting to relate that nine of the crew perished.

But none of the other six local papers referred to the loss of any of the crew. We can speculate as to what happened and why. So here's an alternative version.

The Thetis passed Carrigaholt on the Clare side of the Estuary in the middle of the navigable channel at which point they turned south towards the Kerry coast. Younghusband was heading for Beale, which is where the Estuary turns to the southeast. Beale is the first place on the Kerry side of the Estuary where a landing could be attempted. But in 1834, it was also the last place on the Estuary where Younghusband could safely unload the contraband tobacco. A Coast Guard station had been established just around the corner beyond Beale Point where the Shannon turns. Younghusband could not risk been seen by Coast Guard when he was so far off course. To be safe from prying eyes, he had to anchor to the west of the reef at Beale Bar. He also had to stay clear of the shallow water off Lisadooneen Point. *The Thetis* had no more a few hundred yards of safe water in which to anchor. This would be a risky manoeuvre for a ship even today with an engine. A strong wind may not have been the problem that caused *the Thetis* to run aground. As mentioned earlier, lack of wind could have posed an even greater risk for sailing ships. Either way, wind or no wind, anchoring off Beale Bar was and still is a dangerous manoeuvre and should always be avoided.

The Thetis had to pass Carrighaholt on the Clare side before making the turn for Beale so as to clear the west end of the reef. A ship on the east of Beale Bar is likely to have been spotted by the Coastguard. By then, D'Ombrain had established 160 coastguard stations around the Irish coast with twelve men at each station. D'Ombrain had discovered that the extent of tobacco smuggling throughout the country was enormous and he was determined to break up the combines. There were reports that a tobacco cartel was operating in Kerry with some prominent merchants and landed gentry involved. The Coastguard Station in Beale was situated behind the sandhills but near the highest of the dunes. The top of this dune provided a perfect vantage point to watch west and east along the Estuary. *The Thetis* dropped anchor just off Beale where there was a road leading away from the beach. This road provided access to the population centres of Kerry in Tralee and Killarney where the tobacco was to be sold.

The crew of *the Thetis* launched the rowing boat and loaded the bales of tobacco on to the small boat. Some of the crew brought the rowing boat ashore. They were met on the shore by the people who were the next link in the distribution chain of the contraband tobacco.

But, while this was happening, *the Thetis* ran aground on Beale Bar. If Younghusband had anchored west of Beale Bar, a westerly wind and an incoming tide would push the ship onto the reef. While she was moored, the anchor dragged. The ship was fully loaded with timber so the strain on the anchor was at its maximum. The main anchor was calculated at 1/500 of the ship's weight so the anchor would need to be at least half a ton in weight. But by deck-loading the ship close to the bow an anchor of this size would not have been able to hold *the Thetis* against the wind and the incoming tide.. Raising the anchor would be very difficult with only half the crew on board.

Kerry County Council have erected a sign with this image showing the Thetis on Beale Bar. (Credit - Philip Armstrong).

The ship would need to have the bow directly over the anchor on the seabed to free it. Lifting the anchor to secure it on board would be almost impossible with the wind and the tide pushing against the ship.

If *the Thetis* had been deck loaded with timber then control of the ship was much more difficult. If the anchor pulled free from the seabed, there was little the crew could have done to prevent disaster. Many of the crew weren't on board *the Thetis* so trying to unfurl the sails and pull in the anchor and chain proved too much for the few men left on board. As the crew lost control, the current pushed the ship backwards onto the reef. What is beyond dispute was that *the Thetis* was now in serious trouble. Looking at what remains of *the Thetis* on the beach, the starboard side of the ship had been stove in when the ship hit the rocks. Perhaps, though it is too easy to make flimsy evidence fit a pre-conceived version of that disastrous day in October 1834.

Returning to the facts that night in November 1834 - *the Thetis* was stuck fast on the reef. But, and this is critical to the story, while *the Thetis* was on the rocks, she couldn't sink as the sea was too shallow. Even on the highest of spring tides, it is most unlikely that *the Thetis* could sink while stuck on the reef.

When this happens, the best advice for the crew was to stay on board and hope the incoming tide refloated the ship. It might be uncomfortable with the ship being rocked back and forward but an experienced crew would know what to do. If the two masts did come crashing down, there might be casualties, but this is most unlikely to have happened. Why would the masts collapse? The bowsprit was probably still intact and indeed part of it is still visible on Beale beach 180 years later.

None of this explains the version that stated that nine of the crew drowned. *The Thetis* was aground within a few hundred yards of the shore on the low tide. There was help arriving from local people. There was an occupied house almost on the shore, the ruins of which are still there today so the ship would have been seen to be in trouble. There has been significant coastal erosion in this area since 1834. The Ordnance Survey sheets of the time show the shore was 50 m closer to the Bar than it is at present. One or two of the crew might have fallen off the ship when it went aground, but nine men drowning when their ship was still upright so close to the shore is unlikely. Joseph Younghusband and the remaining crew were able to make their way from the stricken *Thetis* to the shore.

In W.G.Kingston's book *The Three Midshipmen*, from 1873, there is a graphic fictional description of a ship driven ground off the Greek coast. He describes how a line might be sent ashore.

"Some people are collecting on the heights, and will soon be down on the beach," he exclaimed. "Hold on till they come, my lads, and we may be able to send a line on shore." This exhortation was not unnecessary, for the seas rolling in constantly struck the vessel with such terrific force, that it appeared she could not possibly hold together, while two or three men, who had incautiously relaxed their hold, were washed overboard and drowned. A beaker or small cask was in the meantime got ready with a line secured to it. The most important object was to form a communication with the shore. It was

evident that if a hawser could once be carried between the ship and the beach, the crew might be dragged along it and be saved.
– The Three Midship Men, W.G.Kingston.

Perhaps the crew hoped that the rising tide would lift the ship off the reef. However that didn't happen as *the Thetis* never again left Beale. Later, after the arrival of the Customs officials, *the Thetis* was still stuck fast on Beale Bar. It's strange that no effort was made to off load the last of the tobacco before the Customs Officials arrived. Were they interrupted by the sudden arrival of the Coastguard? So perhaps the speculation that Younghusband and his crew had already removed most of the tobacco is not so fanciful after all.

Beale Bar dries out completely at times and it is possible to walk along the reef on the lowest of tides. The tidal range on this part of the Estuary is over 5m, and *the Thetis* drew about 3.5 m so there was a real possibility that she might have refloated. Unless the ship had capsized on her side (beam ends) or had been holed below the water line with the initial impact the ship should have floated on the incoming tide.

We are concerned to state that 44 half-bales of contraband tobacco have been found on board the Thetis by Lt.Triphook and crew of the 'Hamilton', Revenue Cruiser, and Mr. Dexter, Chief Officer and men of the Coastguard at Beale. The greater part was secreted in the seamen's berths, and five bales among the cargo (timber). The crew have been marched in custody from Tarbert to Tralee gaol, to abide the usual investigation. The vessel herself will, it is feared become a complete wreck, but the cargo is safe.
– Saunders News-Letter, Tuesday 9 December 1834.

This report was also printed in the *Tralee Mercury* of December 10th. Later, the *Western Herald* of the 18[th of] December reported that the quantity of tobacco seized by Joseph Dexter was 1,795 lbs. The precise nature of the figure would seem to lend credibility to this account. Contraband American tobacco sold for 1s /lb wholesale, making *the Thetis* cargo worth €2,230 at present day values. And in the term 'half bale' there's a clue about how organised the smugglers were. Each of the bales weighed 41lbs (1,795 lbs divided by 44 bales) which could be easily handled by crew members. A small bale of hay (36" x 20" x 14") weighs about the same. Presumably, these half-bales of tobacco were roughly similar in size. Full bales of 81lbs would be too heavy and awkward to offload from *the Thetis* on to small rowing boats. Two men would be needed on both the ship and the boat.

Dr Sean Whitney wrote a detailed account of the tobacco industry in Ireland during his tenure at the University of Limerick. He dealt extensively with tobacco smuggling in the early 1800s. Dr Whitney's describes how the tobacco was cured:

Most of the tobacco used in Ireland was grown in Virginia, Kentucky and Maryland. Prior to shipment the tobacco leaves would have been cured by plantation owners... The leaves were packed into hogsheads capable of holding 1000 pounds by weight.

According to Dr Whitney, hogsheads were 48" high and 30" diameter holding 1000 lb each. Is this what Joseph Dexter of the Coastguard found on board *the Thetis?* That would change the narrative completely. The calculation of 44 half bales would mean there were 22,000 lbs of cured tobacco which could be worth least 1s/pound. This means that at 1830s prices, there was £1,100 worth of tobacco on board. Using a inflation multiplier from 1830 to the present day of 25, this puts a value of €27,500 on the smuggled tobacco. However, the weight of that type of bale would make them unmanageable while being off loaded onto small rowing boats.

And, of course, none of the above estimates include tobacco that might have been off-loaded in Kilbaha on the Clare coast before crossing the estuary to Beale. Nor does this include tobacco that was taken off *the Thetis* in Beale before the Customs Officials arrived. Younghusband and his crew must surely have made every effort to dispose of as much of the evidence as possible. Could this be just fanciful nonsense? Probably.

Dr Whitney again:

Those involved in smuggling continued to ply their trade into the 1830s when they began to develop new methods to thwart the efforts of the Coast Guard. Local agents would sell tickets initially to tobacco merchants, general merchants and landed proprietors and later to the less well-off including farmers and artisans, who would often form themselves into 'combines' to purchase a part share in a tobacco bale. Any remaining tickets went on public sale in local businesses, a fact D'Ombraine was personally witness to.

The Custom Officials boarded *the Thetis* a few days after she went aground on Beale Bar. They were able to access not only the seamen's berths but also the lower deck where the cargo of timber was located. Obviously, this means *the Thetis* didn't sink. The Customs Officials were led by Lt. Triphook and Joseph Dexter, Coast Guard's Chief Officer. Dexter wasn't happy at the reporting of events. He wrote to the *Limerick Chronicle* in December 1834 to complain that he was described as 'merely an assistant on the occasion'. In his letter, he sets out precisely 'the facts connected with the above seizure' detailing the exact amounts of tobacco found on board.

An early revenue steamer in Foynes.

It has been suggested that it might be the *SS Garryowen* but Brian Goggin's *Waterways and Means* has a sketch plan of the ship that disproves that theory.

To the Editor of the Limerick Chronicle
Coast Guard Station Beale Dec. 11 (1834)

Sir, Having seen a short paragraph in your paper of the 6th of this month relative to a seizure made on board the Thetis now a wreck on Beale Bar and making me merely an assistant on the occasion, I beg leave to state the facts connected with the above seizure. On 9 pm on Monday the 1st I commenced for the second time a close search on board the above brig assisted by James Harris chief boatman and crew from the Cashin station when we discovered 15 quarter bales of tobacco. At 1.30 am the tide beginning to flow round the brig we all quit he: again when the tide ebbed so as to allow our going again on board, we (the same party) discovered 17 quarter bales more besides several small packages making in all 24. On Tuesday Mr. Triphook commander of the Hamilton revenue cruiser arrived and on the following Thursday when discharging part of the brig's cargo (timber) two of the Cashin crew discovered seven more quarter bales of tobacco which will appear to be the only part of the seizure at which Mr. Triphook was present. I would not take the trouble to explain but I feel it would be injurious to my character as an officer to be only an assistant on my own station in making an seizure under the above peculiar circumstances.by you inserting this statement in your valuable journal you will doing me an act of justice by which I shall ever feel obliged. I am Sir, your obedient servant.

Joseph Dexter,
Chief Officer.

Tellingly, Dexter made no reference to the loss of any crewmen. The Custom Officials weren't the only people who turned up at the scene. Three prominent local magistrates Francis Crosbie, William Sandes and William Hickey were also in Beale. They were complimented in a newspaper report for their concern.

Although residing at a considerable distance from the scene of destruction (they) were promptly on the spot making every effort to secure life and property......

> This morning, at an early hour, the body of a new-born infant was found in the River Lee, near the marshes of Ballyard, in the neighbourhood of this town, supposed with too much reason, to have been deliberately drowned by some unnatural monster!
>
> On Saturday night, the Brig Thetis, on her passage from Quebec to Limerick, struck on Beal Bar, and was lost, when, melancholy to relate, nine of the crew perished. The Captain, Younghusband, and five men, were rescued from a similar fate, and received every possible attention and relief, which humanity could dictate from the officer who commands the Coast Guard at Beal.— Messrs. Crosbie, Sandes and Hickie, magistrates, also rendered every possible assistance in saving life and property on the unfortunate occasion.
>
> The Lord Lieutenant has offered £50 reward for the persons concerned in the violation of Mary M'Carthy's person, near Listowel, a few nights since.

A report from a local newspaper Kerry Evening Post.

William Sandes had been appointed High Sheriff of Kerry in 1828. He lived in Sallow Glen, near Tarbert. The Crosbie and Sandes were local landed gentry and were connected by marriage. Sandes is a name long associated with the area of Moyvane which is often referred to as Newtownsandes. The Crosbie family had large estates and houses in Ardfert, Ballyheigue and Ballylongford. William Hickey lived in Kilelton House, near Ballylongford. The three men were local landed gentry with a great deal of power and authority in North Kerry.

The report states that the magistrates rendered every possible assistance, but what does this mean? Wading out in the sea or rowing out to *the Thetis* which was stuck on Beale Bar to help rescue the crew? And why were they there at the time *the Thetis* went aground? Perhaps they had been waiting for the ship to arrive? Three prominent magistrates were on the scene in Beale but were unable to record the names of the nine crewmen who drowned. It doesn't make sense from start to finish. This extract from Dr. Whitney provides a more plausible explanation for their presence.

An Indication that members of the land'd gentry were involved arose in County Kerry in which Colonel Crosbie, of Ballyheigue Castle, and governor of that county, whom a former water guard officer believed to be a smuggler, came under increased suspicion when a cave was discovered on the lawn of his residence in clear view of any inhabitants.
– Commissioners of Inquiry 1824, W. Lloyd to James Dombraine, appendix 52. PhD Thesis, Dr Sean Whitney.

The Colonel Crosbie referred to is James Crosbie of Ardfert. He had a son, Francis Crosbie, who was thirty years old in 1834. Presumably, this is the Francis Crosbie who was on Beale Beach in November 1834. Colonel Crosbie was having serious financial problems which were worsened by the collapse in agricultural prices in Ireland.

The three magistrates may have other reasons for being on Beale apart from 'making every effort to secure life and property'. The facts as we know them, are as follows. There was contraband tobacco on board *the Thetis* that day in November 1834 when the ship went aground on Beale Bar. Tobacco combines were known to be operating in the North Kerry area. The son of one the largest landowners in the area who was suspected to be a smuggler happened to be at the scene on the fateful day. If somebody was going to pay large sums of money to bring the contraband tobacco to Ireland, then it would make sense to have somebody you could trust as the cargo was brought ashore. The tobacco had to be transported from the beach and distributed to the other members of the cartel. Monies had to be collected for the tobacco. And who better to look after this than your own son?

The Thetis foundered on Beale bar on the 30th November 1834, but ten days later a local newspaper report makes no mention of the death of nine sailors. Nine local men drowning should have been a major tragedy. It would have been extensively reported in the *Limerick Chronicle* or any one of another five local newspapers but there is no mention of such an incident in these papers apart from that small note in the *Kerry Evening Post*. On Monday December 8th 1834 the *Caledonian Mercury Edinburgh* reported that *'the Thetis* is still on shore and discharging her cargo nearly all of which is expected to be saved'. There is no report of loss of life.

In 1834, the deaths of these sailors would have been the subject of a major inquiry and an inquest. There is no record of any inquiry. A report from a newspaper, *Saunders News-Letter*, eight days after the grounding says that *the Thetis* is still on shore. But 'the cargo is safe'. Younghusband would never have abandoned his ship when there was the slightest chance that the tide would lift *the Thetis* off the reef. The idea that the crew would leave their nine dead shipmates behind is not credible either. The bodies of some of the drowned sailors would have washed ashore on the incoming tide. On this six-kilometre-long stretch of beach, some of the missing men would have been recovered. There is no record of this happening. One account said that the crewmen were buried at Lisadooneen Point just beside Beale Beach. In 1834, this couldn't have happened. It would never have been allowed. The remains of the drowned seamen would have to be brought back to their families for burial. If that couldn't be done there was a graveyard two miles away in Kilconly and another larger graveyard in Lislaughtin, Ballylongford. People weren't buried on the edge of a cliff in 1834. There may have been graves on Lisadooneen Point but they must have been from an earlier shipwreck.

There may be another explanation. Up until 1815, the Royal Navy had used a system of 'pressment' or press gangs to fill the ranks aboard their ships. Men could be forcibly pressed into service on Navy ships for a period of five years. However, since 1815 the Royal Navy had been reduced from 145,000 sailors down to just over 15,000 men. Press gangs no longer operated, apart from one group of men, convicted smugglers. They could be pressed into service in the Royal Navy instead of a prison sentence. Some of the crew could find themselves forced to serve in The British Navy. Henry Baynham quotes the case of William and Edward Burnett, who were given five years' naval service in 1832, as punishment for being found with contraband spirits. Given the huge reduction in Royal Navy personnel some of *the Thetis* crew may even have experienced life in the Navy. Some may have been deserters although that is unlikely as the Navy was greatly reducing numbers. Either way, just blending into the rural countryside of North Kerry must have seemed a better option than running the risk of five dangerous years in the British Royal Navy.

Eleven years earlier in 1823, *the Phoebe* was reported by the *Connaught Journal* to have been wrecked off Glenbeigh in County Kerry. *The Phoebe* was on route from St. Johns, Brunswick in Canada when on the 3rd March 1823, she went aground. All the crew were reported as being saved. The captain was called Fullerton Key. Richard Mahony a local magistrate from Portmagee, Co. Kerry carried out a full investigation. This contrasts sharply with the lack of investigation into the circumstances surrounding the wrecking of *the Thetis*. So did Younghusband stay with his ship? Almost certainly, he did because a newspaper report say he didn't abandon *the Thetis*. For all his faults, Joseph Younghusband was an experienced captain and knew what was expected of him. A report from the *Limerick Chronicle* confirms this.

The quantity of tobacco was only 44 quarter bales of about 25 lbs each and those were brought on board surreptitiously by some of the crew who are in custody. The Master, it appears knew nothing of the matter and has not been taken up as erroneously stated: he is at wreck endeavouring to save the property.
– Limerick Chronicle.

If the Master, Joseph Younghusband, knew nothing of the matter, why was *the Thetis* on the Kerry side of the Estuary? The ship was there because the captain brought her there. Shortly afterwards, it was reported that Younghusband and five of his crew were arrested in Tarbert by Joseph Dexter of the local Coastguard. Perhaps they were waiting there for a Marconi Bians to take them back to Limerick. Or as mentioned earlier, this could refer to the general Tarbert area rather than the village of Tarbert. The other nine men were reported drowned or more than likely fled the scene. Younghusband and the remaining crew were marched to Tralee. Of course, they were, it's only 26 miles from Tarbert to Tralee. They would easily be there in two days. This seems farcical and makes no sense. If you're going to make up a story, at least make it credible. And a small report in the *Limerick Chronicle* seems to confirm this:

She lies inside Beale Bar and as the weather has now moderated we trust the loss may not be great. The master continued fast by the vessel all through when most of the crew had deserted her.
– Limerick Chronicle, 1834.

Joseph Younghusband and the remaining crew were tried in a Tralee Court that conveniently neglected to keep a record of the proceedings. Nobody went to prison. There is one report that mentions a fine but even that isn't substantiated by any records. Joseph Younghusband was apparently sacked by Francis Spaight, or was he? And what happened to *the Thetis*? Presumably, the timber cargo was offloaded and either confiscated by the Revenue or perhaps delivered to the owner Frank Spaight in Limerick. Because remember, he didn't know anything about the tobacco smuggling.

The Coast Guard did enjoy early successes if one measures it by the number of seizures it made in the early 1820s. Prior to its establishment, only nine vessels carrying smuggled tobacco had been seized between 1800 and 1819. In the first five years of the Coast Guard's existence, twenty-two ships were seized. A total of 1,768,818 pounds of tobacco was seized by the authorities between 1821 and 1823 which, while encouraging to those battling smuggling, also served to highlight the magnitude of the problem.
– Dr. Sean Whitney.

Final Voyage of the Thetis

Continuing the account related to George Greaney in 1898, grandson of Dick Greaney who was an apprentice seaman on *the Thetis* in 1834.

Looking back sixty five years, I still can't believe Captain Joseph made such a terrible mistake that night in November 1834 in the Shannon Estuary. They said, later, he was intoxicated but Younghusband wasn't drunk that night. He never drank much at sea always waiting till he came ashore. He argued with Connors about going too close to the Kerry shore. I was on the helm right beside them and heard every word. They were beholden to the Kerry cartel and had to get the tobacco ashore at Beale. Connors knew the dangers; he argued that the anchor wouldn't hold against the wind and tide. Captain Joe wouldn't hear of it. 'If you don't like it get off the ship with the tobacco and don't come back.' Connors stormed off and jumped ship as soon as he could. I couldn't believe it when he turned up ten years later, on The Native with Younghusband. Anyway, once the anchor pulled it was all over. With the running tide and the south westerly wind against us, we had no chance. Younghusband had made one of the biggest mistakes of his life. The crash when we hit the reef was sickening. We were stuck fast. I knew, immediately, we were in trouble and wouldn't be coming off the rocks. The sails were mostly stowed so the masts held. But we were on our side and we were finished. Water was pouring into the hold and we knew our brave little ship was gone. Connors and nine of the lads were on shore loading the tobacco onto the local carts. They were supposed to come back to get the rest of the tobacco but instead they had to come back through the surf to get us off The Thetis. We had some job getting Captain Joe off his ship. He had to leave but he made sure he was the last off. He stood on the shore looking at the Thetis being slowly battered to pieces. Was there a tear in his eye? Whatever the case, he straightened himself and said, 'are we all here'? And we were. 'So', he said, 'those who want to go or have to go, leave now. The Customs will be here soon and you don't have to face them. We'll say you were lost. Spaight will take his cargo off the ship so why should you give up your freedom? Now quickly, go home to your families, see you in Limerick, lads. We'll sail again.'

Today, only the timber beams of *the Thetis* are still on the beach, 187 years later. The cargo was presumably removed and brought to Limerick. But the rest of the valuable material on board had to go somewhere. There were miles of ropes (halyards, lines, and sheets) on board *the Thetis*. There were acres of canvas which could be reused as sails or covers in farmyards. Blocks, tackles, brass fittings and valuable timber in the hull of the ship were all left behind on Beale Bar. The two main masts, 30m high were extremely valuable. After all ships were sailing to Quebec to obtain these long timbers. But nothing of *the Thetis* has survived locally, nor is there any anecdotal evidence of looting of the stricken ship.

Inevitably though, some local breaking of the ship must have happened. When it became obvious that *the Thetis* could not be refloated and when the cargo had been removed, the ship was abandoned on Beale Bar. But in calm weather, the wreck was easily accessible by small rowing boats so if Spaight didn't want any more to do with his ship then an impoverished local population could find a use for everything that was left behind. One further question remains …. what happened to the anchor? *The Thetis* had a 3m high, 9cwt (450 kg) cast iron, bower anchor which has not been accounted for. If the anchor was used when the contraband tobacco was being off loaded, then could it be still lying on the seabed. The timber hull and masts of these ships will rot away, and the canvas and ropes will deteriorate but the cast iron anchor will last much longer. If Younghusband realised that *the Thetis* was stuck on Beale Bar why would he leave the anchor connected to the ship? With an incoming tide, the anchor could have pulled the bow of *the Thetis* under the water. Younghusband would have no choice but to release the anchor chain. There could be a large cast iron anchor waiting to be discovered a couple of hundred yards from the beach just west of Beale Bar. Spaight may have wanted to distance himself from *the Thetis*, particularly if he and Sam Evans had managed to convince the insurers to pay out. But there is a fawning newspaper report of the time that contradicts the suggestion that *the Thetis* was ransacked.

> *We have heard with much pleasure that the conduct of the peasantry in the neighbourhood of Beale in this county where the Thetis was driven ashore has been most praiseworthy. They never evinced the slightest disposition to plunder but have throughout shewn an anxious wish to save for the owners any property that came on shore. The Magistrates Messrs Sandes, Crosbie and Hickey have been constantly in attendance and expressed themselves at all times ready to render any assistance in their power and that it was not their intention to seek any remuneration by lodging salvage claims. Such disinterested conduct is highly creditable in those Gentlemen.*
> *– Limerick Chronicle.*

This is almost certainly nonsense. 'Those Gentlemen' were not standing on Beale Beach inhaling the sea air for the good of their health. If they were so concerned about the events unfolding in front of them, why didn't they record the drowning of nine members of the crew? They were Magistrates, Officers of the Crown, so were duty bound to investigate the reported loss of life. It is doubtful, also that the 'conduct of the peasantry' was so well behaved that they could have cared about the owner's property. And of course, why would they or should they care about the owner's property? The same article in the *Limerick Chronicle* contrasts the wrecking of *the Thetis* with the fate of the *City of London* (this should probably refer to the *City of Limerick*), a ship which was plundered by the local population in Ballybunion when it ran aground in 1833.

The Thetis crew are not listed anywhere in any records. Apart from Younghusband, there are no names available of any of the eighteen men who were on board *The Thetis* in November 1834. However, there are records of crews on some other of Spaight's ships. *The Native* had Peter Morgan, John Hugham, Thomas Henright, Stephen Kormill, Robert Taylor, and Pat O'Connor on board in 1843. Patrick O'Brien, Pat Behane (apprentice), George Burns (apprentice), Pat Cusack,

John Gorman (cook), William Griffiths, John Palmer (passenger) and Michael Behane died when *the Francis Spaight* sank in 1835.

In 1846, another ship called *Francis Spaight* was lost en route to Manila from London. A whaleboat took fifteen crew members off the stricken vessel. Henry Patterson (captain), David Evans, Thomas Haydon, Henry Hiate, and Magnus Smith had survived the 1835 tragedy but all were lost on *the Francis Spaight II* in 1846. James Robertson survived in 1846 by staying on board *the Francis Spaight*. As these ships were owned by Francis Spaight, it is possible that some of these men were also on *the Thetis*. Many of the names were not common in Ireland. Hiate and Kormill were almost certainly not from Limerick.

And that was that. Well, no it wasn't, not by a long shot. Francis Spaight and Joseph Younghusband's story was only halfway through. After being supposedly sacked by Francis Spaight, Joseph Younghusband made at least one more trip to Quebec but this time it was from Liverpool. In 1835, he was the captain on *the Plata* which left Liverpool on the 3rd April and arrived in Quebec on May 14th. He, then, disappears from the records for several years.

Charles Bianconi was opening up Ireland with his network of horse drawn carriages based in Clonmel. By 1832, he had 300 horses and his cars, known as Bians were travelling 1800 miles per day. His network included a route than ran from Limerick to Tralee and onto Caherciveen. At one time, Tarbert was a stop on a Bianconi route. Perhaps Younghusband and his remaining crew used the Bians when travelling from Tarbert to Tralee for the court case and then back home to Limerick in late 1834.

EIGHT
The Native (February 1843)

Francis Spaight appointed John White as captain of *The Native* in 1843. *The Native* was a schooner making regular trips from Limerick to London. This was a strange appointment as White had no experience of captaining an ocean-going sailing vessel. He had spent several years as skipper of *the Dover Castle* making regular trips on the Shannon from Limerick to Kilrush. *The Dover Castle* delivered goods to Kilrush during weekdays. On weekends, it brought day trippers to Kilrush who were then brought by road to Kilkee for a day at the seaside. This was a five-hour trip, in sheltered waters, down the Estuary one day and back the next day.

In September 1840 the company gave Captain White the use of the Dover Castle for a pleasure trip, accompanied by the Limerick Musical Academy; he was to receive the profits from that trip.... The Limerick Shipping Company's Sunday trips to the mouth of the estuary provided a few hours ashore at Kilrush in the afternoon, but drunkenness and rowdiness dissuaded the more sober customers. –Limerick Chronicle

The *Limerick Chronicle* reported that on one of the excursions 'several of the passengers were so violent and intemperate from a too liberal indulgence in Bacchanalian orgies that the captain and crew could not maintain order……'

Eventually, due to dwindling demand for its services, *the Dover Castle* was sold. When John White took up his position as captain on *the Native*, he needed a first mate who knew about ocean sailing to make up for his own lack of experience. He chose Joseph Younghusband whose employment with Francis Spaight had supposedly been terminated after the loss of *the Thetis* in 1834. Spaight claimed afterwards that he was not aware that Younghusband was on board one of his ships.

Spaight didn't have a record of sacking his captains, even when ships were lost. Both Gormans continued working for Spaight after surviving more serious shipwrecks. Experienced transatlantic captains were hard to find so why would Spaight fire Younghusband? In fact, it is notable how many men captained *the Thetis* across the Atlantic, some of whom made only one trip indicating how arduous and difficult the job was. It doesn't stand up to much scrutiny that Spaight wouldn't know that Younghusband was on board. After all, how was Younghusband to be paid on board *the Native?*

Wouldn't Sam Evans, Francis Spaight's clerk, have to be aware of the men on board all the ships? Crew manifests had to be lodged with the authorities both in Limerick and in the away ports. So perhaps Younghusband was kept in reserve or served as first mate on board Spaight's ships since 1834.

But by then since the early 1830s, Francis Spaight had perfected the art of not knowing. Tobacco smuggling, overcrowding of his ships, drownings and court cases were some of the things that Spaight managed not to know about.

Either way, in 1843 nine years after *the Thetis* sinking, White and Younghusband were in charge of *The Native* on a voyage from Limerick to London. For some reason, that is very difficult to explain, they concocted a scheme to scuttle *The Native* and steal the cargo. The following is an account of the whole sorry episode taken from Brian Goggin's *History of Inland Waterways* called the 'Non-Return of The Native':

After about three weeks off Ryde, The Native left Motherbank on 6 February 1843 and, that night, anchored near Jack in the Basket, the outer marker at the west side of the channel to Lymington... The sails were furled and the crew were sent ashore to buy potatoes, although they already had enough. John White, before removing the crew, cleared the run of the vessel, and bored two of the holes half through himself [with an auger]. Three more holes were bored by Younghusband and Connors, while the captain was absent with the remainder of the crew at Lymington. They were all plugged up, and the plugs removed next day, when the vessel was at sea. White and Younghusband had stolen the cargo entrusted to them and had done so unknown to anyone except several boatmen and the grocers on the Isle of Wight - and their own crew, who must have known that something was up. Then they sank their schooner, in full view of the crews of two pilot boats. After that they returned to Limerick.

Jack in the Basket.

'Unknown to anyone' except at least twenty others, *the Native* had been anchored off the Isle of Wight for three weeks before stopping again near Lymington in The Solent. Lymington and its surrounding coastline was renowned as a haven for smugglers. The proximity of the island meant that contraband from the continent could be dropped there and moved onto the mainland. Were White and Younghusband involved in smuggling? Even if they were, it doesn't explain the strangeness of the whole affair.

What is striking about this, apart from the recklessness, is the amount of detail about the incident that's available. The crew members are known. Pat O'Connor, sometimes referred to as Connors, was the most prominent of the crew: Connors took his meals with Younghusband and White, which was unusual, as ordinary seamen did not normally eat at the captain's table. He was implicated in the scuttling but got away scot-free. He was involved in the whole illegal enterprise but somehow disappeared before he could be brought to justice.

Another passage from *Irish Waterways History* lists some members of the crew of *the Native*:

The others of the crew were Peter Morgan [also given as Magan, Megan, Migan and Mignon], who had joined in London, John Hugham [also given as Huhan], "a Scotchman, belonging to Salcoats", Thomas Henright, who was lost overboard in a gale, Stephen Kormill, an apprentice, and Robert Taylor, who "was to work his passage to Limerick".
– Irish Waterways History.

This is in contrast with the lack of information about the crew of *the Thetis*. Younghusband and White were arrested after returning to Limerick and were sent back to London to face trial. The details of the trial are recorded and readily available. Francis Spaight travelled to London and gave evidence at their trial. They both pleaded guilty and offered no justification for their actions. The two men were convicted and transported for life to Australia. The Judge told the court that:

.... the prisoners, they had been convicted, upon their own confession, of one of the most serious charges, short of taking away life, known to the law, viz that of destroying at sea a ship. Unfortunately for the purpose of defrauding the owners". Because ships' officers had "so many opportunities of inflicting very severe injury upon the public" with little chance of detection, a very severe punishment was warranted. He therefore sentenced White and Younghusband to be "severally transported beyond the seas for the terms of their natural lives".
– *On Monday 15 May 1843,* Mr Justice Coltman.

Transportation of Britain's convicts to her colonies was adopted as the answer to the problem of overcrowded jails. Capital punishment was considered to be excessive for many of the lesser crimes. Sending men and women to Australia became a common punishment handed out for both major and petty crimes. Transportation to Britain's colonies continued from the seventeenth century until well into the nineteenth century. On 26th August 1843, Joseph Younghusband was transported on *the Maitland* to Norfolk Island, arriving there on 7th February 1844.

Norfolk Island.

Norfolk Island is a small, isolated island of 14 square miles located in the Pacific Ocean, about 1,400 kms east of Australia. The journey from Britain to the penal colony on Norfolk Island took about five months. This must have been a humiliating and demoralising experience for Younghusband. Instead of captaining the ship, he was confined below decks in crowded accommodation often kept in shackles.

In 1824 the British government instructed the Governor of New South Wales, Thomas Brisbane, to reopen Norfolk Island as a place to send "the worst description of convicts". Its remoteness, previously seen as a disadvantage, was now viewed as an asset for the detention of recalcitrant male prisoners. The convicts detained have long been assumed to be hardcore recidivists, or 'doubly-convicted capital respites' – that is, men transported to Australia who committed fresh crimes in the colony for which they were sentenced to death, but were spared the gallows on condition of life on Norfolk Island.
– Wikipedia

Why was Joseph Younghusband treated so harshly? He doesn't seem to fit any of the above descriptions. He wasn't the captain of *the Native* and yet he received the much tougher and ultimately fatal sentence. He died six months after arriving on Norfolk Island, aged forty one. His prison record gives no details about his appearance or how he died; except to say he was married with four children.

Joseph Younghusband's prison record from Libraries Tasmania.

John William White was transported on *the Anson* on 23 September 1843, reaching Van Diemen's Land, or Tasmania as it now known, on the 4th February, 1844. White's prison record has more detail about his physical appearance. He was 5'8" with 'a fresh complexion'.

John William White's prison file from Libraries Tasmania.

His wife and family followed him in 1846. According to his prison file, White was released on a ticket of leave in March 1852 which meant he could work while remining in Tasmania. White was able to recover some sense of a normal family life and spent many years working in Hobart as a shipping clerk. He died in Hobart, Tasmania, on the 15th November 1867, at the relatively young age of 59. White's prison record also says that he and Younghusband scuttled *the Native* to obtain the insurance money. White and Younghusband didn't own the ship so they couldn't have benefited from any insurance claim. There was only one person who could have claimed the insurance money and that was the owner of *the Native*, Francis Spaight.

There are so many unanswered questions about it. They both must have harboured a grudge against Francis Spaight for the way they were treated by him. However, Younghusband supposedly hadn't had any dealings with Spaight since the grounding of *the Thetis* in 1834, some nine years prior to the sinking of *the Native*. So that doesn't explain it either. But Younghusband and White returned to Limerick where they knew Spaight was waiting for them. There must be more to this than meets the eye. Both men had families in Limerick, so that partly explains why they took such a risk. Perhaps, they thought they could make their way to America and out of the reach of the law and Francis Spaight. This entire episode makes no sense.

The detailed records of the scuttling of *the Native* contrasts with the lack of detail surrounding *the Thetis*. The episode of *the Nature* shows that accidents at sea in the 1800s were treated with the utmost seriousness. Inquiries were held, people were arrested, court cases were prosecuted and the guilty were held accountable. In all cases, deaths were scrupulously recorded. And the many newspapers of the time reported all the details including names and ages of those who lost their lives. But in the case of wrecking of *the Thetis* nothing happened. This is despite a lost ship, contraband tobacco, and the reported death of nine sailors. This adds to the frustration of not being able to put names to the extraordinary crewmen who sailed the Atlantic on board *the Thetis*, 190 years ago.

Francis Spaight, at least, provides us with a happy ending to the story. He continued in the shipping business, buying bigger and bigger boats. Of course, being a modern man, he moved with the times and bought steamships.

Advertisement
Emigration to America. Important notice to Emigrants

Francis Spaight has purchased this season a splendid new oak ship named the Jane Black the largest vessel ever in the port of Limerick, passengers will therefore have on board this large vessel all they can desire for their comfort and accommodation.
– 1842 Wednesday 30th March (Limerick Chronicle).

The advertisement listed Spaight's ships, including *The Jane Black* which weighed 1300 tons and was captained by Timothy Gorman. The other ships referred to in this notice were *the Borneo*, captained by P. O'Donnell, *the Governor* captained by Daniel Gorman, *the Thetis II* (700 tons) with Daniel Ross as captain and *the Bryan Abbs* (600 tons), with J. Hugill on board. Of course, we know Frank Spaight's claims about the weight of his ships should be taken with a grain of salt. Still, it seems that the ships were getting bigger. And across the Atlantic, despite the introduction of lighthouses, the dangers to shipping in the Gulf of St. Lawrence continued.

List of Vessels Wrecked coming to Quebec during the Season 1841, with the number of Lives lost.

On the 14th May, The brig Breeze, from Limerick, was wrecked on the Island of Scatari, with 160 passengers, who were saved, but lost their baggage and provisions. The brig Minstrel, from Limerick, was wrecked on Red Island Reef, with 141 passengers and 15 crew, only eight persons survived. On the 26th September, the Amanda, from Limerick, was wrecked on Little Metis Point, with 18 crew and 39 passengers, 41 lives were lost.
– The Shiplist (website).

Francis Spaight retired to his estate at Derry Castle in the 1850s handing the running of the business over to his son, James. But 190 years later the wreck of *the Thetis* still lies on Beale Strand as a silent monument to the events of 1834. Sometimes, when *the Thetis* is uncovered by the sea and the sand, I stand in the middle of the remains of this small ship. I try to imagine what it must have been like to be packed into this space with over 200 other people for six weeks as *the Thetis* crossed the Atlantic. *The Thetis* did her best, despite all the disadvantages of size and design, to bring over a thousand people from Ireland to the Americas in search of a better life. Of course, it's not my ship but can I be forgiven for thinking that I might own just a small piece of the story of *the Thetis*?

Well no, not according to the voice in my mind when I walk on Beale Strand - *not bad*, it says *but you forgot to mention the tears, the sadness, the terror, the friendships, the songs and the laughter of 217 passengers and crew, as we shared forty two days of our lives crossing the Atlantic on board the Thetis in 1834.*

Engineer and author of this book, Paul O'Dowd, standing by the wreck of the Thetis, on Beale Beach, Kerry.

Epilogue

An epilogue is where everything gets wrapped up and unanswered questions get dealt with. But not here, as there are too many details lost and will probably never be recovered. In 1834, most of the passengers on *the Thetis* had some money and many had family connections in the Americas. Even though the ships were completely unsuitable, people were healthier than immigrants during the famine years and so were better able to endure the Atlantic crossings. While this book is not about the Great Famine of the 1840s, the quarantine island of Grosse Ile plays a significant part in the story of *the Thetis*. And yet the story of Grosse Ile is largely forgotten in the Ireland of the 21st century.

It wasn't until 1865 that shipowners were required to lodge passenger lists with the authorities. Even so there were many ships from other Irish ports making the passage across the Atlantic where passenger and crew names were diligently recorded. This was not the case in Limerick.

Then there were the ordinary sailors on *the Thetis* who brought hundreds of immigrants safely across the Atlantic and we don't know who they were. Men who risked their lives day and night on these long voyages, but were often insulted and disregarded by the passengers on board. So much so, that their names have been forgotten.

If nine men drowned that day in November 1834, shouldn't we know their names? Or perhaps we should accept that they didn't die that night, but rather lived on to sail again and probably die on another one of Francis Spaight's ill-fated ships.

We know that in the 1830s, a tobacco smuggling combine was operating in the North Kerry area. And members of the landed gentry were involved. It is also certain that Beale Beach was a landing point for this tobacco. However, another one of the unanswered questions is how much of this did Francis Spaight know?

If Spaight wasn't a member of the combine, how was the smuggled tobacco paid for in Canada? Who was the paymaster? If members of the landed gentry, Sandes, Crosbie and Hickey were part of the tobacco smuggling combine then Frank Spaight must have known about it.

But what are we to make of Joseph Younghusband? Here was a young man in his early thirties who brought hundreds of people to Canada in a small ship which was not fit for purpose. He left a wife and young children behind in Limerick as he captained *the Thetis* across 2500 kms of the Atlantic Ocean. Younghusband had to deal with the chronic overcrowding of *the Thetis* on the way to Quebec and the dangerous overloading of the ship on the return leg to Limerick. He managed to keep his passengers and crew largely free from disease which caused so many deaths on other ships. It was Younghusband who had to suffer the consequences of the wrecking of *the Thetis* even though he was probably carrying out his employer's instructions. He was accused of being intoxicated on board *the Thetis*, which, given the circumstances surrounding the events that night in November 1834, was almost certainly not the case. This seems to have been an effort to cover up the real reason behind the wrecking of *the Thetis*.

Joseph Younghusband was a complex individual who was not averse to some shady behaviour, as his part in the scuttling of *the Native* shows. However, while his co-conspirator John White was sentenced to a relatively benign captivity in Tasmania, (Van Diemen's Land), Younghusband was sent to the much harsher regime on Norfolk Island, and when there paid the ultimate price, dying in his early forties. Perhaps not a hero or a villain but a bit of both.

> *Coincidentally, the "Quebec" (Q) signal flag, called the "Yellow Jack", is a simple yellow flag that was historically used to signal quarantine signalling. But it now means the exact opposite. It signals that the ship is healthy. In the phonetic alphabet, Quebec stands for Q, so the connection to Gross Ile and Quebec is maybe coincidental.*

References

Henry, W. (2009). *The Coffin Ship: Wreck of the Brig St. John.* Mercier Press.

Dana, R. (1996). *Two Years Before The Mast.* Wordsworth American Classic.

Gallagher, T. (1986). *Paddy's Lament.*

Duffy, S. (1997). *Atlas of Irish History.* Gill Books.

Whyte, R. (1847). *Famine Ship Diary.*

Keane, M.C. (2021). *Crosbies of Cork, Kerry.* MCKeane.

Knight, R. (2022). *Convoys.* Yale University Press.

Murphy, B. and Vlahou, V. (2018). *Adrift.* De Capo Press.

Sobell, D. *Longitude.* Fourth Estate.

Golding, W. (1980). *Rites of Passage.* Faber and Faber.

Junger, S. (1997). *Perfect Storm.* Fourth Estate.

Foster, R.E. (1988). *Modern Ireland.* The Penguin Press.

Royal Observatory Greenwich Guidebook.

Chancellor, J. (1984). *John Chancellor (Art).* David & Charles.

Kealy, R. *Historical Jewels (Limerick).*

Whitney, S. (2019). *The Irish Tobacco Business.* University of Limerick.

Clarke Historical Library, CMU.

The Shipslist, Website.

Guillett, E.C. (1963). *The Great Migration*. University of Toronto Press.

Hewson, M. *Emigration to the North America, Colonies Limerick 1841. Reminiscences of Old Limerick*.

Barrie, D. (2014). *Sextant*. William Collins.

Allan, P.K. (2017). *The Captain's Nephew*. Penmore Press.

Higgett, R. *Finding Longitude*. Collins.

Cunliffe, T. (1989). *Celestial Navigation*. Fernhurst Books.

Marryat, F. *Mr. Midshipman Easy*.

Ingles, H.D. (1838). *Ireland in 1834*. Whittaker & Co.

Ballantyne, R.M. (1857). *Coral Island*. Puffin Classics.

Doyle, M. *Hint of Emigration Canada*.

London, J. *A True Tale Retold*.

Strickland, C.P. *The Backwoods of Canada*. Project Gutenberg.

Moodie, S. *Roughing it in the Bush*. George P. Putnam.

Steel, D. (1794). *Element and Practice Rigging*.

Melville, H. *Redburn: His First Voyage*. Penguin Classics.

Fitzroy, R. *The Beagle*.

Harvey, D.C. *The Wreck of The Astraea*.

Baynham, H. *Smugglers*.

Goggin, B. *Non Return of The Native*. Inland Waterways History.

About the Author

Paul O'Dowd is a Dubliner who has lived in North Kerry with his wife Anne for over 40 years. An engineering graduate from UCD (1970), he has long been interested in the history of navigation. A shipwreck on a beach near O'Dowd's family home has always intrigued him. He describes himself as a 'fair weather sailor', but despite this he has undertaken many maritime adventures, including sailing across the Mediterranean. Previous adventures include a sail to Greece and Croatia, a sail from Spain to Dingle and from Cork to the Scilly Isles. He has explored most of the south west coast of Ireland in his own yacht which is based in Fenit, Co.Kerry. *Final Voyage of the Thetis* is Paul O'Dowd's first book.

Acknowledgements

I must thank many people who helped me with this book. Starting with my sisters who kept me from straying too far off the path of historical accuracy.

Samira Kassis from the Lebanon is a friend of ours who has lived in Limerick for a number of years. Two years ago, Samira insisted after reading an early outline that I finish the book. She has also contributed photographs to the book and then used her extensive knowledge of Limerick City to track down the Francis Spaight monument. So, thanks to Samira, Frank and Sara for their support.

Thanks also, to my neighbours in North Kerry, Michael Flahive, Donal Liston and John Hennessey who regularly enquired about how the book was progressing. Their encouragement was very important. Thanks also to Gary Mulvihill, in my office, who had to put with time spent in the 1830s instead of the 21st century.

In Listowel, one day, I asked Billy Keane to recommend an editor. He said the ideal person lived just around the corner from my office and so I was introduced to Jeremy Murphy and his team at JM Agency. So, thanks, Billy.

Jeremy, Sarah and Parvathi, who with unfailing patience and professional courtesy made this book a reality. Jeremy introduced me to the world of book publishing and all its intricacies. I am very grateful to everybody at the JM Agency.

Thank you to the staff in the libraries in Tralee and Limerick who were always so courteous and helpful. I received encouraging emails from Rachael Kealy who has written about the Limerick merchant class including the Spaight family in her series 'Historical Jewels of Limerick'.

And thank you to Dr. Sean Whitney who took the time to reply to my requests for information about tobacco smuggling. Arlene White of the Kilrush Historical Society provided me with helpful contributions.

And of course, thanks to my own family, my wife Anne, my grown-up children, Frank and Kate and my daughter-in-law, Patricia who have had to listen to stories about the 1830s for the last three years.